1981

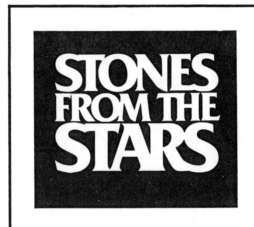

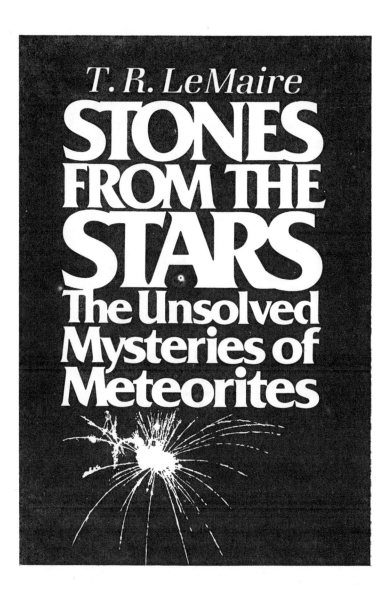

T. R. LeMaire

STONES FROM THE STARS

The Unsolved Mysteries of Meteorites

PRENTICE-HALL, INC., Englewood Cliffs, N.J.

To Christie and Stephen

Stones From the Stars: The Unsolved Mysteries of Meteorites
by T. R. LeMaire
Copyright © 1980 by Theodore Rogers LeMaire
Printed in the United States of America
Prentice-Hall International, Inc., London
Prentice-Hall of Australia, Pty. Ltd., Sydney
Prentice-Hall of Canada, Ltd., Toronto
Prentice-Hall of India Private Ltd., New Delhi
Prentice-Hall of Japan, Inc., Tokyo
Prentice-Hall of Southeast Asia Pte. Ltd., Singapore
Whitehall Books Limited, Wellington, New Zealand
10 9 8 7 6 5 4 3 2 1

Library of Congress Cataloging in Publication Data

LeMaire, Theodore Rogers.
    Stones from the stars.

    Bibliography: p.
    Includes index.
    1. Meteorites. 2. Meteorite craters. I. Title.
QB755.L39 1980   523.5'1   79-21158
ISBN 0-13-846881-8

CONTENTS

Part One

INTERPLANETARY VISITORS

Chapter One

LOST CITY, OKLAHOMA

Fireball Trajectories ☆ *Photographic Puzzles* ☆
Electromagnetic Sounds

On the evening of January 3, 1970, a ball of fire burst from the skies over northeastern Oklahoma. Glaring incandescent in a furnace of friction as it penetrated our atmosphere, and announcing its arrival from space with sonic booms, the cosmic intruder hurtled toward Earth at speeds of over 30,000 miles per hour.

Meteors remain one of nature's most majestic if short-lived phenomena. Accompanied by brilliant pyrotechnics and startling sound effects, they can stun any audience, leaving most observers with breath enough only to gasp. For nine dazzling seconds the javelin of light astonished witnesses below, illuminating the winter landscape for 400 miles like a full moon. The fireball flamed out some 12 miles above ground.

Moments later, at 8:14 P.M., a charcoal-black lump of stone made its mystifying rendezvous with Planet Earth.

Any meteorite recovered from a fireball or from a previous unreported landing is of enormous interest to science, for although we may dispatch rocket-powered vehicles on voyages extending millions of miles, none is yet programmed to send back any mementos. True, we possess a few Moon rocks, but these were collected in our own spacial backyard. Until one of our spacecraft makes a round trip, ferrying back a bit of Martian or Mercurian real estate to Earth, these objects of stone or iron remain our only solid contact with deep space.

Scientists prize meteorites as means of studying "the geology of space" in the search for clues to the birth of our Earth, asteroids, and the solar system itself. But strangely enough, the color, brilliance, and velocity of a fireball are seldom accurate guides to the nature of the object creating the aerial extravaganza. The chunk of space matter displaying such splendors may be stone, iron, or a combination of the two, and often the heavenly show is costume only. When later retrieved on the ground as a rare and valuable meteorite by some lucky Earthling, the object may look disappointingly small, perhaps being only the size of a loaf of bread.

The hurtling halo emblazoning the Treysa, Germany, fireball of 1916 measured about 400 feet across, but the central meteorite itself was only 15 inches. "The flare was bigger than a football stadium," wrote a reporter, "but the meteorite was the size of the football." Scientists believe that the Nuremberg meteor, torching the skies above Germany and Czechoslovakia 57 years later, was only a marble at heart. The fiery flare coursing over southern England and northern Ireland in April 1969 measured 660 feet across, while the recovered meteorite weighed but 12 pounds.

Not every meteorite will fit into an ordinary rock garden, though. After the sonic blasts from a wildly spiraling space-object cracked plaster in a Norton County, Kansas, courthouse in February 1948, a huge stone weighing more than a ton was recovered from a field—becoming the world's largest stone specimen ever discovered from an eyewitnessed fireball. That record lasted a quarter-century, until in March 1976, northeastern China was visited by a space stone weighing almost 2 tons.

But even these stony giants are pygmies compared with one behemoth iron still lying imbedded where it landed in the Kalahari Desert of South-West Africa (Namibia). Broad as an elephant's back, the 60-ton Hoba West meteorite is the reigning world champion. Our planet, however, has been visited by even more monstrous meteorites thousands of times heavier, which have self-destructed, blasting out enormous craters many miles wide.

Nevertheless, few fireballs actually drop chunks of stone or iron, and little evidence normally remains of their spectacular appearance. Numerous fireballs inexplicably detonate in midair, leaving behind a cloudy "dust trail" which high-altitude winds soon waft away. Others apparently burn themselves out in the fiery friction of our atmosphere, merely winking out like an electric light switched off; or disappear at full speed over the horizon in mysterious defiance of our local earthly "laws" governing gravity and aerodynamics.

To date, only about 2,000 meteorites are available for laboratory analysis. But even the trajectory of an incoming meteorite can be revealing. By mathematically backtracking the flight path, astronomers can retrace the object's orbit around the Sun for fresh insights into the asteroid belt between Mars and Jupiter.

For years, investigators had tried unsuccessfully to capture on film that blazing streak marking a meteorite's course through our atmosphere. Any number of low altitude fireballs and high altitude meteors had been photographed, but none had accommodated scientists by dropping a chunk of stone or iron on film. But when a meteor burst from the winter skies over Oklahoma on the evening of January 3, 1970, a camera trap was waiting—clusters of automatic aerial cameras known as the Prairie Network.

Purpose of the Prairie Network: to recover a meteorite. Established by the Smithsonian Astrophysical Observatory of Cambridge, Massachusetts, and financed by NASA, the network consisted of 16 unmanned stations, housed in concrete sheds. Cameras stood trained on the nightly heavens of the Great Plains states from Oklahoma north to the Dakotas and from Kansas east to Illinois. This region was chosen because of freedom from obstructing hills, and also because of Kansas's cosmic reputation. (For reasons unknown, meteorites have fallen upon Kansas with curious frequency.) The cameras had now been waiting 5 years.

Television and radio in Oklahoma and neighboring states reported the fireball's arrival, for the amount of light given off by these objects is almost incredible. In an early June morning of 1964, a fireball whooshing over Saginaw, Texas, southwest of Fort Worth, gave off a light so intense that it created a false dawn, causing some street lamps, which automatically turn off with daylight, to extinguish at 3:30 A.M. In 1954, one astronomer calculated the candlepower from a fireball flashing across California, Nevada, and Arizona at 70 billion watts—enough, he figured, to illuminate the entire United States that evening. Colors bursting from these speeding, luminous space-objects are no less spectacular, often displaying a rainbow effect. They may burn blowtorch blue, blast-furnace orange, lightning white. In 1968, fishermen in the Gulf of Mexico reported a fireball that began as a greenish glow, then turned to red, and finally blue. In Germany, an American tourist described the April 1975 space visitor as "dazzling greenish with orange sparks falling down."

Thus it was that the camera network manager, at home in Nebraska on January 3, first learned of the fireball while watching the evening news on television. Immediately he made operational a plan for processing the film from the two stations nearest the fireball's trajectory. The negative rolls were picked up and flown to Lincoln, Nebraska, for special processing, then air-dispatched to Cambridge for computer analysis. Scientists there factored into their equations all available data for locating the meteorite's probable impact point. This information included the fireball's speed, its angle of descent through the sky, estimated size and weight, air resistance and the velocity of high altitude winds, gusting that night at 150 miles per hour.

Cambridge calculations suggested that this particular meteorite had fallen in Cherokee County, east of the Gibson Reservoir in the scrub-oak foothills of the Ozark Mountains; not the easiest area to search. Eight inches of snow covered the ground. Scientists placed the point of impact approximately three miles east of a small farming community with the disconcerting name of Lost City. The computer specialists telephoned instructions back to the network manager in Nebraska, who then drove 450 miles in his pickup truck to Oklahoma and onto the snow-filled roads near Lost City.

He later admitted that he had had little hope of finding the precious meteorite, and had planned to ask local farmers to watch for any strange, dark-colored rock come spring when the snows melted. But luck rode with him. The manager spotted the sky-prize, black against the white snow, sitting smack in the middle of the road. He leaped from his truck and retrieved the 23-pound cosmic visitor. The 9-inch stone was a rough teardrop in shape and known as an "oriented" meteorite, meaning that it had kept one face turned toward the ground, boring straight through our atmosphere like a miniature space capsule.

Thus was found the Lost City meteorite, which *Sky and Telescope* magazine reported to be ". . . the first recovered in a search that was guided by the photographically determined trajectory of a meteor through our atmosphere." Oklahoma's cosmic visitor was about to go down in history, with scientists never suspecting that their precious pictures—a strip of light on a 9-inch-wide negative—would only compound the many mysteries of fireballs and meteorites.

Scientists were ecstatic at their cosmic detective work in recovering the meteorite, but slightly miffed that the specimen had landed about a half-mile from their calculated landing spot. Technical minds are not accustomed to such error. Rechecking their work, they found that the "mistake" lay not so much in their computations as in the meteor's bizarre behavior.

Velocities far greater than the Oklahoma visitor are known, of course. The Saginaw flare of 1964 was clocked by radar at 73,300 M.P.H. Such a fast-moving object, if impervious to heat and if able to stay aloft, would cross the continental United States in less than three minutes. The Oklahoma fireball had entered our atmosphere from space at a fairly normal 31,200 M.P.H. Suddenly, however, its speed dropped to 7,500— *this* was unprecedented. Furthermore, it had flared out at an altitude of only twelve miles.

These twin perplexities had been fed into the computer. Two other factors were far more mystifying. In guarded language, reports explained how the fireball "had departed appreciably from a linear path." In other words, it had made an abrupt change in direction. Some force, other than 150 M.P.H. winds aloft, had caused it to swerve from its original course. Another puzzlement: The fireball had shown "an unexpected lateral acceleration." In everyday language: it had not only swerved in midair, but speeded up at the same time. (No wonder calculations were off by a full half-mile!)

There is a sequel to the unexplained antics of the Lost City fireball. On January 18, a farmer searching for a stray calf picked up a small dark stone half-buried in the frozen sod. This turned out to be a 10-ounce fragment that had separated from the parent body, landing some 2,000 yards from the impact point. Later, investigators found still a third piece nearby. In commenting on this scattering, NASA stated, ". . . the two smaller fragments of the meteorite appear to have been affected by aerodynamic lift forces during their flight." Or—a "lift" far from anything expected. Like manned gliders, the fragments evidently had leveled off in flight and flown an additional 2,000 yards.

The Lost City fireball was no maverick: for decades, trained astronomers and other professional sky observers have been sighting flaming objects that behave most strangely, but the majority of their accounts must be sought in journals aimed at specialized branches of science. Few reports are readily available to the general public because scientific papers, bristling with figures and couched in esoteric jargon, are usually too uninviting for the popular press.

But with technical talk pruned away, the kind and variety of sightings is astonishing. The following anomalies are taken from an assortment of technical publications. Reports are not chosen on the basis of extraordinary sightings, for one is as astonishing as the next. They merely suggest the diversity of unexplained phenomena that fireballs have generated during the past three-quarters of a century.

When the Indiana fireball of July 12, 1952, passed near Lafayette, a professor of aeronautical engineering from Purdue University was among the thousands of startled observers. He watched the luminous mass turn color, from faint yellow to brilliant blue. With the light change, he heard a distinct hissing sound. Puzzled, he plotted the fireball's course, relative to his vantage point, on a chart. His conclusion: to have seen and heard the meteor simultaneously was impossible—according to known laws of physics.

Theoretically, the hissing should have arrived much later because of the vast difference in velocity between sound waves (moving at about 1,100 *feet* per second) and light rays, traveling at some 186,000 *miles* per second. At a distance of 20 miles from the trajectory, he should have heard the object about a minute and a half after its change in color; not at the same time. According to his observation, the fireball was transmitting sound waves at the speed of light—an "impossibility."

Was the engineer hearing things? Not at all. Scores of similar reports have been collected from around the world. Even at a distance of 100 miles, reliable witnesses have reported seeing high altitude meteors and hearing them simultaneously.

Low altitude fireballs frequently produce sonic booms of terrifying intensity, shattering windows, cracking plaster. But despite the power of such thunderings, these acoustics are but atmospheric disturbances, created when any high-speed object, be it fireball or jet plane, breaks the sound barrier. The human ear, however, is not the only instrument capable of picking up fireball sounds. With the object's passage, radio sets have been infected with snappings and clickings. As was radio station KSFO, San Francisco, in 1963.

In the early evening of November 7, a fireball was sighted traveling west over downtown San Francisco. The object continued out to sea, where it flared brightly before disappearing. One witness reported, "Just as I stuck my head out of the window, I heard a funny sound, like 'Whoosh'—like the sound of a Roman candle shooting off—and I saw this object sail over the trees . . . a bright blue flame about twice the size of a dinner plate [and] shooting sparks in all directions."

The fireball's passage took several seconds, followed by sonic booms which witnesses likened to "thunder" and "train rumblings." So far, so good. But unlike most fireballs, which burst so suddenly against eye and ear, this one had given advance notice. Three minutes before arriving, it alerted San Francisco radio listeners of Station KSFO.

At 6:13 P.M., a *Beep-Beep-Beep* noise at 1-second intervals cut into a disc-jockey program. For the benefit of his audience, the puzzled announcer ad-libbed, "This is KSFO, the station with the mysterious beep." The engineer tried to filter out the interference. Unsuccessful, he switched on the tape recorder. Three minutes later, at 6:16, the fireball made its debut.

Later the tape playback was analyzed by audio technicians, who concluded that the beeps were caused by an equipment malfunction within the station. They attributed the interference to "an accidental triggering of an 880-cycle oscillator, used to mark the time on the hour." Yet the beeps had started thirteen minutes *after* the hour.

Perhaps the experts exonerated the fireball too hastily. While it was going *whoosh* over San Francisco—and while KSFO was going *Beep-Beep-Beep*—two power stations of the Pacific Gas and Electric Company were going on the blink. In the suburbs of San Bruno and Westlake, lights were dimming as 4,000-watt fuses cut out.

Do fireballs transmit electro-phonic beeps? Were these happenings around coastal San Francisco only a stack of coincidences three layers deep? Well, make that four. Two years following the KSFO "malfunction," a huge ball of fire was reported falling on the Pacific shore, near Anacortes, Washington. It went *Beep-Beep-Beep*.

And back in 1948, scientists bouncing radar beams off high-altitude meteors reported, "Some bright meteors emitted a signal independent of transmitted radar pulses." In other words, the objects were beaming back radar signals of their own!

Suspecting some undefined type of electro-magnetic force, engineers refer to such sounds as "electro-phonic noises." The name may be appropriate, yet it hardly explains the numerous other fireballs that have been observed performing intricate maneuvers in our private skies, whether depositing a meteorite or not. Scientists, in dutifully recording these authentic yet weird findings, often revert to the term *anomaly*. In essence they are saying, "Here is a verified observation: we can't explain it because it just doesn't fit into our present realm of knowledge." Not only do some fireballs speed up and change course, as photographed at Lost City, but others perform loops and spirals, dive and climb—even run zigzag courses. Many such events date back a half-century and more, years before the tabloids and entertainment media became entranced with the likes of flying saucers and UFOs. Nor are reports limited to the

United States. From Austria comes the 1932 account of a fireball splitting in half—the smaller segment making a wide U-turn in the sky and heading away in the opposite direction from its parent mass.

Equally baffling are those others that, in total disrespect for celestial mechanics, amble across our skies all too leisurely, staying aloft with no visible means of support. Because of the heat generated by air friction, a foreign body entering our atmosphere has a life expectancy measured in seconds. When a space-rock does glow to visibility, it already is burning at temperatures of 5,000° F. and up, hotter than a blast furnace melting iron. The object must either be consumed by flames, or plummet to the ground. The Lost City flare lasted only nine seconds; a twenty-second span is most unusual. Nevertheless, reports of fireballs flashing about the skies for several minutes—even for as long as an hour—are not unknown.

One spectacular example, known as "The Great Meteor of March 24, 1933," performed marvelous acrobatics in the skies over New Mexico and adjoining states for almost a half minute. The show opened with a fireball streaking across the heavens on a southerly course. It turned west on a zigzag track. Then, some witnesses maintained, it separated into two parts, which traveled side by side. At one point, the glowing object was enveloped in a ball of luminous gas from which leaped a gigantic orange spark.

A year later, residents of Idaho and Nebraska enjoyed an aerial extravaganza lasting far longer. Illuminating the landscape for hundreds of miles, a fiery object traveled at a leisurely 8,000 M.P.H., rather than the orbital speed of 18,000 M.P.H. needed for sustained flight. The flaming body then separated—not into 2, but into 3 segments. The trio continued in a V-formation, the largest leading. Occasionally members of the troop would switch positions. Estimated flight time: 6 minutes.

This stellar performance, however, seems downright hurried when compared with events of June 23, 1950, high over the Gulf of Mexico. The following account is taken from a study conducted by a most experienced member of the American Meteor Society who interviewed numerous witnesses ashore and dug through logbooks of merchant vessels made available to him by the U.S. Navy Hydrographic Office.

Second Officer R. M. Richardson, aboard the American vessel S.S. *Polaris Sailor,* logged the meteor's first sighting at 10:39 P.M. The officer observed a "meteor falling rapidly, developing in size and brilliance until reaching an altitude of 11 degrees, when it disintegrated with a shower of fire." Elapsed time, according to the logbook: four seconds.

10:40: Aboard the Danish ship S.S. *Martin Carl,* Second Officer M. Thoneson watched "a meteor of 1 minute and 2 seconds duration . . . It left a bright trail, first straight and later becoming zigzagged."

10:41: Aboard the American S.S. *Dartmouth,* Chief Officer M. L. Frankline saw "a meteor . . . or phenomenon . . . It traveled in a westerly direction to an altitude of about 20 degrees, where it appeared to make a complete loop, and change course to the north."

10:45: Third Officer K. E. Rudolph recorded in the logbook of the American S.S. *Pan-Virginia,* "What appeared to be a bright meteor was observed . . . altitude 30 degrees. The phenomenon fell to an altitude of about 20 degrees, where it swerved to the right and upward, then curved again."

10:50: The master of the American S.S. *Alcoa Cavalier* witnessed a "strange phenomenon . . . It seems like an object zigzagging across the sky."

An incoming meteorite, of course, has no power supply. No guidance system. No way of maneuvering in space. The object's trajectory is rigidly governed by its initial direction and cosmic velocity, plus the known effects of gravity, air pressure, and friction. It is not a guided missile. Legitimate meteors, of course, lack wings, engines, jets, or propeller; yet, according to these logbook reports by merchant marine officers, the Gulf of Mexico fireball remained aloft for eleven minutes. Its prolonged flight perplexed the Meteor Society investigator. Especially puzzling to him was the object's trajectory. "It was not approximately straight," he wrote, "but curved in a most peculiar manner or spiral."

The "oddball" behavior of these fireballs can be rationalized by assuming that very natural if undetected forces were at work, and in time science will solve the riddle. The argument, unfortunately, is much too fragile when contrasted with an amazing series of sightings, made in 1913 in lower Canada and eastern United States. That phenomenon still has Space Age scientists frowning into their computers.

Chapter Two

THE CANADIAN FIREBALL PROCESSION OF 1913

Natural Meteors or Alien Spacecraft?

Despite the controversy over their origin, scientists are in agreement that the dozens of luminous objects that paraded across the night skies of south-central Canada and northeastern United States were real—not the product of any mass hallucination or any kind of hoax. Impeccable were the credentials for the primary witness to the Procession, who also served as its chronicler.

Professor C. A. Chant of Toronto University was a member of the Royal Astronomical Society of Canada, and editor of its official journal. After interviewing numerous witnesses, Professor Chant wrote a full report, substantiated by other investigators during and since the event.

The "meteors" were first spotted in the winter skies near Regina, capital of Saskatchewan Province.

"At about 9:05 on the evening in question," Professor Chant's colorful and articulate account in the astronomical journal stated in part, "there suddenly appeared in the northwestern sky a fiery red body which quickly grew larger as it came nearer, and which was then seen to be followed by a long tail. Some observers state that the body was single, some that it was composed of two distinct parts, and others that there were three parts and each followed by a long tail . . . In the streaming of the tail, it resembled a rocket, but unlike the rocket, the body showed no indication of dropping to earth. On the contrary it moved forward on a perfectly horizontal path with peculiar, majestic, dignified deliberation; and continuing on its course, without the least apparent sinking towards the earth. . . ."

Professor Chant described the object's color as fiery red or golden yellow. He wrote that its brilliance resembled "the open door of a furnace," "the illumination of a searchlight," "a stream of sparks blown away from a chimney by a strong wind." His report continued:

"Before the astonishment aroused by this first meteor had subsided, other bodies were seen coming from the northwest... Onward they moved, at the same deliberate pace, in twos and threes or fours, with tails streaming behind ... They all traversed the same path and were heading for the same point in the south-eastern sky ... Several [witnesses] report that near the middle of the great Procession was a fine large star without a tail, and that a similar body brought up the rear. ..."

The mysterious glowing orbs were traveling southeast. They crossed into Minnesota from Canada and then over the Great Lakes, passing about 25 miles south of Toronto. From Saskatchewan to the Atlantic coast, scores of spectators marveled at the celestial extravaganza, their descriptions differing only in details. At Alpena, Michigan, the Weather Bureau reported: "A group of 10 brilliant meteors, one following the other in a straight line, appeared in the northwest and passed over the city in an easterly direction. Each gave off a faint reddish trail of light. Shortly another group ... appeared ... apparently traveling in the same direction." In Gainsville, New York, one eyewitness wrote to the local newspaper: "A shower of burning meteors crossed the sky to the eastward, leaving a trail of fire and smoke that hid the stars for several minutes." Some observers suggested that the Procession may have stretched across the sky for a hundred miles or more.

Passing directly over Buffalo, New York, the "shower of stars" cut across the corner of New York State and Pennsylvania, moving out to sea near Sandy Hook, New Jersey. Subsequent ship sightings placed them next in the vicinity of Bermuda, and ultimately in the South Atlantic below the equator, about 200 miles off the Brazilian coast.

It is difficult to estimate the actual number of bodies making up the star-spangled show. Some spectators saw only one or two solitary bodies, while others nearby counted dozens. The greatest number was mentioned by a Trenton, New Jersey, high-school student watching with binoculars. He saw "about ten groups in all and each group ... consisted of from twenty to forty meteors."

An accurate census evidently was impossible because, like Fourth of July sparklers, members of the heavenly troop persisted in bursting into frenzies of fiery fragments. A weather observer in Upper Michigan saw a large fireball mount above the horizon: "... it separated into 2 bodies, each about half the size of the Moon: when these reached the zenith, they burst into a great number of small bodies which continued to the south-eastern horizon."

Another witness reported seeing "a meteor break into about fifty pieces like the explosion of a skyrocket." A group of New Jersey church-goers was treated to an especially dazzling show. According to a local newspaper account, "As the congregation was leaving the musical service last evening at Watchung, the members were astonished to see 7 stars shooting across the sky . . . They followed each other in rapid succession, and 5 of them passed out of the line of vision, but the remaining 2 seemed to explode like skyrockets, and burst into a thousand tiny stars. . . ."

But even though Professor Chant identified the bodies as natural meteors, they were not "shooting stars." Rather than plunging toward the ground like most meteors, these bursting, flaring objects held a constant altitude, as Professor Chant remarked, while maintaining a kind of formation in flight. "One was following directly in the path of the other," stated a Boyne City, Michigan, resident. "Like they were strung on a wire" was an expression used by two different observers. At Fenlon Falls, Ontario, one man likened the Procession's leisurely movement to "the angular speed of a flying crow."

The Canadian Fireball Procession of 1913 maintained an almost horizontal path, neither gaining nor losing altitude, for some 5,600 miles, or almost one-quarter revolution of our planet before disappearing. "There is nothing in the annals of meteoritics," exclaimed one investigator, "that can compare with this path-length."

With the glare of air friction, meteors plunging to Earth usually become visible about 90 miles above the globe. The atmosphere quickly brakes their burning speed, however, cooling the surface, quenching flames; and the object disappears from view at an average height of some 30 miles. Professor Chant's calculations placed the Procession at an altitude of not more than 30 miles—remarkably low for a sustained trajectory of any natural bodies and close to the bottom line of visibility. Furthermore, despite their low velocity—Professor Chant estimated that the Toronto aerial display lasted 3.3 minutes—the bodies apparently produced what we now call sonic booms. (Back in 1913, of course, practically no jargon of the Air Age had entered people's vocabularies.) Professor Chant compared the Procession's noise to "a carriage passing over rough roads or a bridge." Another witness described "a rumble like that of a railway train or a distant waterfall." (No one mentioned beeps.) Because no reports later returned from across the Atlantic Ocean, many authorities assume that the Procession never reached Africa, but fell into the sea.

What was the Canadian Fireball Procession of 1913? A cluster of natural space-rocks that gravity dragged down from the atmosphere to an unmarked grave in the South Atlantic? One fact is certain: the massed display of flaming bodies, traversing the sky above the Great Lakes with militarylike precision, was not born of this Earth, because back in 1913 the aviation industry was just getting off the ground.

Only ten years had passed since the Wright Brothers were airborne in their marvelous flying machine at Kitty Hawk, North Carolina. When Canada entered the Air Age in 1909, a flimsy apparatus of silk and bamboo took to the air at Baddeck Bay, Nova Scotia, for a half-mile hop. In 1909, Louis Blériot was proud indeed to be the first aviator across the English Channel with a then-formidable flight of 31 miles. In the United States, several months later, Glenn H. Curtiss (who had recently been awarded this country's first pilot's license) won a $10,000 grand prize for his 150-mile trip from Albany to New York City, averaging a breathtaking 50 miles an hour.

Another decade passed, however, before the broad Atlantic was spanned. In 1919, a U.S. Navy seaplane pushed from Newfoundland to the Azores, while a British lighter-than-air floated through a round trip to Mineola, New York, in 7 days—not counting time for stopovers. These ocean crossings still fell some 2,000 miles short of the Procession's astonishing flight-track of three years earlier.

Professor Chant was the first to propose that the "shower of stars" within the Canadian Fireball Procession consisted of swarms of meteors clutched by the Earth's gravitational pull and circling the globe as natural satellites—a myriad of mini-moons. He assigned these satellites the necessary velocity of 18,000 M.P.H. to keep them aloft, and positioned them in a neat circular orbit. His theory explained their horizontal track, level with the horizon, and constant altitude above the ground.

His calculated height of not more than 30 miles placed them in the denser layers of our atmosphere, causing them to burn red hot. According to the professor's theory, during their prolonged flight between

Saskatchewan and the coast of Brazil, they were totally consumed, with final remnants disappearing just below the equator. But other scientific minds soon criticized his explanation. One investigator, analyzing distances traveled by the Procession in the given time, arrived at a speed of less than half the 18,000 M.P.H. necessary to keep the satellites airborne. He further insisted that unless the objects were of great tonnage, they would have burned out long before finishing their 5,600-mile trek.

Another authority flatly rejected Chant's hypothesis, stating that the circular orbit and low altitude made his theory impossible. This scientist argued that our Earth is not a perfect ball: it protrudes many miles at the equator; the entire fleet of "meteors," therefore, would have collided with this bulge before reaching Brazil.

The few scientists still interested in the Procession today retain Chant's basic premise of natural meteors, while altering the mechanics of his explanation. One theory replaces the straight-line parade of objects with a towering swarm of space-rocks, orbiting the Earth in a stack perhaps hundreds of miles high. According to this hypothesis, gradually the "flying gravel pit" lost altitude. As the lower echelons of pellets trespassed into our guardian air, they briefly became luminous in the night—single clusters of shooting stars, each flaring and dying within seconds, yet creating the illusion of a "procession" across the sky.

In 1961, a high-ranking official from NASA's Goddard Space Flight Center offered an even more sophisticated explanation in *Sky and Telescope* magazine. He was looking to aerial displays like the Procession as rationale for the origin of tectites. These beautiful glassy objects in bottle-green, black, and brown are shaped like buttons and tiny dumbbells. Rare to most parts of the world, tectites are highly prized by laboratories, museums, collectors.

The NASA scientist first assured his readers, "Many of the shower's phenomena are illuminated by the physical and engineering studies, especially of satellite re-entry, that have been made in the last few years."

For his theory he then replaced the flying gravel pit with an enormous space-rock traveling around our planet in a grazing orbit, making a complete revolution of the globe once every 91½ minutes. Slowly this huge asteroid started dropping lower, theorized the official, to be blasted by furnace-hot temperatures of atmospheric friction. The melting surface spewed off droplets of molten material that observers identified as shooting stars, and that undoubtedly congealed in the air to fall as tectites. Some of this incandescent matter, however, followed the main orbit, which "constituted the bodies of the shower."

Necessary for the NASA hypothesis were repeated orbits by the meteor-like objects, because the scientist's calculations "failed to show any way in which all members of the shower could disappear in [only] one revolution."

During the intervening 91½ minutes, the magazine article continued, our rotating globe would have turned beneath the orbiting asteroid to slide other portions of North American geography and other observers under the sky-drama. On its next pass around the world, the Procession would be seen in an area southwest of the Regina-Sandy Hook track. "The next revolution, with a period of 91½ minutes," the NASA engineer wrote, "would have carried them over the Middle West, above the populated regions of Nebraska, Iowa, and Missouri." He logically assumed that such a spectacle would have been reported in those states no less than in Saskatchewan and New York, and their brilliant performance featured in local newspapers.

For evidence supporting his theory, the scientist doggedly searched through scores of daily and weekly publications ("practically the entire collection of the Library of Congress"), for write-ups of a meteor shower on the night of February 9, 1913. Strangely, he found nothing, and admitted being "unable to locate a single article referring to the shower." Neither did he report discovery of any gem-precious tectites along the Procession's track.

Will scientists discover some unknown "law" of celestial mechanics to supply the missing pieces for the 1913 aerial puzzle? Or is a ready-made answer now at hand, as was suggested in 1969 in *Watchers of the Sky* by Willy Ley who wrote: "The whole behavior of the meteorites seemed totally strange to people of 1913, but now it can be compared with the behavior of an artificial satellite which has re-entered the atmosphere at a very shallow angle, and is carried halfway around the Earth before impact takes place."

But once again, this rationale doesn't quite fit all the facts. For in his investigation, Professor Chant meticulously plotted on a map the course of the Procession from Regina to Sandy Hook. His ground track is known as the Chant Trace. Unknowingly, he preserved evidence of an astonishing performance by the Procession, a feat unrecognized until now. (See Fig. 1.)

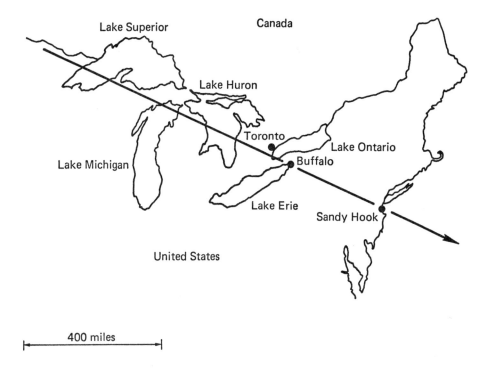

Figure 1. The Chant Trace   The line across the map marks the course of the 1913 Canadian Fireball Procession, as drawn by the Toronto astronomer who meticulously investigated the phenomenon. First sighted near Regina, Saskatchewan (not shown), the Procession traversed the five Great Lakes and headed southeast over the Atlantic at Sandy Hook, New Jersey.

In 1956, the Canadian Fireball Procession became the target of an intensive study by the American Meteor Society as reported in *Meteoritics*. Its investigator, Alexander D. Mebane, was extremely thorough in his 18-month analysis. He corresponded with 615 newspapers in towns located along the Chant Trace, requesting information from their files about events on the night of February 9 and the early morning of February 10, 1913. He interviewed weather bureaus in seven states, inquiring into any unusual observations on that date, questioned local historians and librarians, and advertised in Pennsylvania and New Jersey newspapers for personal accounts of still-living witnesses.

Although overcast skies limited visibility that night, and although results of Mebane's "extensive canvass were somewhat meager," he did succeed in multiplying the marvels of February 9 and 10, reporting that ". . . a certain amount of new evidence has been obtained, which, if it does nothing else, at least definitely serves to rule out . . . [any] spaceship interpretation," which had been proposed over the years by Charles Fort, among others.

Back in 1913, without the precedent of rocket-boosters, space capsules, UFOs, flying saucers, or decaying satellites, several witnesses *did* liken the objects to man-made vehicles. One astronomer compared the Procession to "an express, lighted at night . . . The lights were at different points, one in front, and a rear light, then a succession of lights in the tail." At Marshall, Michigan, one couple mentioned a plane. According to the Marshall *Chronicle,* "Miss Vera Murray and Carl Shnaitman, when returning to the city last night, reported that they had seen an airship going east."

Although Professor Chant's official opinion espoused natural, free-falling bodies, he too specifically mentioned airships: "To most observers the outstanding feature of the phenomenon was the slow majestic motion of the bodies; and . . . the perfect formation which they retained. Many compared them to a fleet of airships, with lights on either side and forward and aft; but airmen will have to practice many years before they will be able to preserve such perfect order. . . ."

According to Mebane, the Procession consisted of free-falling bodies, glowing incandescent from air friction: "members of clusters of closely related fragments originating from the partial or complete disruption of larger bodies." He hitched his conclusion to the personal interpretations of his witnesses. "None of those who saw the Procession at close range described it as other than meteoritic in character . . . [The] more favorably placed observers unanimously recognized the lights as incandescent meteorites. . . ." Mebane disregarded the airship explanation by the Marshall, Michigan, newspaper because of the observers' extreme distance from the Chant Trace.

Later, NASA drew liberally upon Mebane's findings in formulating the theory about a large incandescent asteroid, spewing droplets from its molten surface, as a rationale for the February 9 phenomenon.

During his intensive investigation, however, Mebane made an astonishing discovery, of which he is rightfully proud. Ironically, it seems to testify *against* his own evidence for incandescent meteorites. He learned that there were not only one or two, but rather *three* separate sections to the Canadian Fireball Procession. The first appeared above the Great Lakes region about 9:00 P.M., February 9; the second passed almost unnoticed in the sleepy morning hours of February 10; the third appeared that afternoon.

Mebane is most definite, almost exuberant, about his discovery of the unreported appearance of the follow-up group of fireballs. As reported in *Journal of Meteoritics* Vol. 1, No. 4, 1956, "This fantastic spectacle, a 'Procession' in its fullest glory, was seen directly on the Chant Trace, but some 5 hours later than expected."

Still-living witnesses recalled for Mebane details of the heavenly pageant. Mr. Archer Keller, a resident of the Delaware Water Gap between Pennsylvania and New Jersey, remembered the awe-inspiring spectacle, and in Mebane's presence, "sketched a parade of meteors of gradually decreasing size, each size-class arranged in regular ranks except for the 'go-as-you-please brigade' of red sparks that brought up the rear."

Keller, reported Mebane, described how "the main body, which occupied more space than the full Moon, was definitely dark, surrounded by incandescent light very much like the corona of the Sun in eclipse . . . There were 5 or 6 following meteors, about half the size of the leader, and in perfect alignment; they seemed to revolve as tho' geared together . . . They flared and flamed, belching back exploding sparks and fire . . . Then came another array, about twice as many, of half the size . . . Soon the sky was filled with exploding sparks. . . ."

Mebane found other witnesses to the second section of the Procession, confirming his discovery: a follow-up batch of "meteors" arrived about 2:30 A.M., following the same Chant Trace across the Great Lakes and out to sea near Sandy Hook, New Jersey. And there burn a pair of puzzlements: one that Mebane recognized and one that he did not.

Anyone wishing to fly an airplane (or spacecraft) on a straight course from Regina, Saskatchewan, touching each of the five Great Lakes, must be a pretty fair navigator. A map of the lakes reveals that one line, and only one line, can be drawn to intersect all bodies of water. Three of the lakes—Superior, Michigan, and Huron—present little challenge, because of their size and position. Any number of courses, fanning east from Regina, will cut across their shores. And an easy connection can be had with a fourth lake: either Ontario or Erie, but not *both*. The navigator must thread a geographical needle by clipping the western tip of Lake Ontario and the eastern tip of Lake Erie, and passing directly over Buffalo, New York.

This is the precise course laid down by the Chant Trace. (See Fig. 2.)

If there had been but a *single* Procession, this presumed navigational feat could be dismissed as chance. The objects had to follow *some* course above the surface of Earth: why not one touching the five Great Lakes? The ground track would be a tolerable coincidence, for practically anything can happen once. Twice, suspicions are aroused. Three times, and the verdict must go against coincidence and free-falling bodies.

According to Mebane, the second section of the Procession was followed by a third on the afternoon of February 10. Diligently picking his way through Professor Chant's records, the investigator for the American Meteor Society discovered that "3 dark objects passed west to east . . . repeating the Procession once more on precisely the same course as before."

All three sections of the Canadian Fireball Procession threaded the needle. Yet by following the same ground track across Earth, each section of the Procession had to fly a *different* course through space!

Does the Sun rise and set? No, it's an illusion. The Sun stands still. As solar observers aboard a rotating planet, Earth, it is *we* who are moving, completing a revolution of 360° every twenty-four hours. Continuously, our spinning globe offers a different patch of its surface to a stationary Sun. Thus, for any one part of the world—such as the Great Lakes—to be constantly bathed in sunlight, either our planet must stop rotating or the lamp of heaven, to hang above the Great Lakes, must chase this globe around and around.

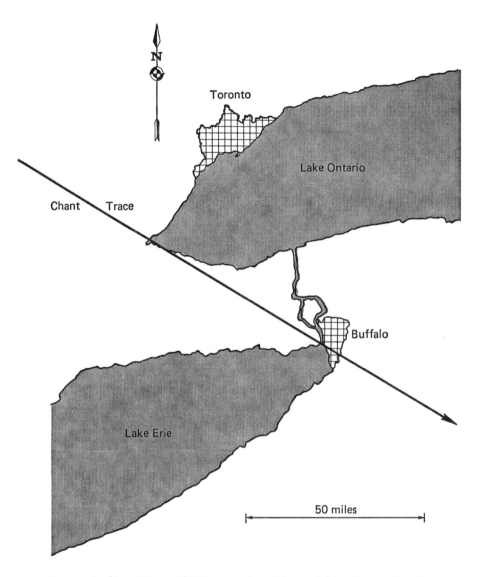

Figure 2. Chant Trace Closeup   Any object wishing to overfly all five Great Lakes on a straight-line course must clip the tips of Lakes Ontario and Erie to do so. This is the precise ground-track followed by each section of the Procession: *three* groups of fiery bodies repeated this unique line across the Earth's surface.

And that, essentially, is how the Canadian Fireball Procession be-haved. During the 5-hour lapse between the first and second sections, as Mebane realized to his consternation, our Earth would have turned away from his "meteors" 75°—in terms of terrestrial geography, some 5,000 miles; offering the space-objects a view, no longer of the Great Lakes but of the Pacific Ocean.

Yet *all three* sections of the Procession threaded the needle. By dupli-cating the Chant Trace, chasing the Lakes, each had to travel a different course through space.

They also had to chase our planet. The Earth is not parked in the solar system but travels some 66,000 miles every hour along its orbital path. Asteroids, fireballs, and meteorites supposedly are not equipped with propulsion systems or guidance devices, yet consider the navi-gational challenge to the "pilot" of the second section. He not only had to locate the Great Lakes, but first intercept Earth—now 330,000 miles down the orbital road. The "navigator" of the third section had to be equally familiar with celestial mechanics.

Little possibility exists that anyone will ever solve the mystery of the Procession. All evidence, pro and con, must remain circumstantial. That is, until we receive direct confirmation by gathering space-rocks in the so-called asteroid belt, believed by many scientists to be the cradle of the meteorites that fall on Earth—but which presents some harrowing di-lemmas all its own.

Chapter Three

COSMIC TIMETABLES

The 3:00 P.M. Peak ☆ *The Late Spring High* ☆ *The Februarians*

Another area of astronomy offers a kind of precedent for the Canadian Fireball Procession. Over the ages, veritable blizzards of meteors have been admired streaming from space. Meteor showers appear on a rather predictable schedule to be named, numbered, and catalogued by astronomers. Two of the most famous are the Perseids and Leonids, which can be watched this year (as every year) around August 12 and November 15, with varying degrees of intensity. Because each star-storm seems to radiate from a single locus in the sky, like water spraying from a nozzle, the showers are named after the background constellations: "Perseids" for the star-group Perseus; "Leonids" for Leo the Lion.

During spectacular years, the sky becomes a fountain of sparks—sufficiently overwhelming, vouched one wag, for those of sinful heart to repent on the spot. Most astonishing was the 1833 Boston drama. One investigator likened the star-storm to "half the number of flakes that one sees during a snow shower." On the day following, a Yankee farmer reckoned how he would look at the sky come evening, to see whether any stars were left up there.

Astronomers have a theory about these celestial outbursts. In its yearly trip around the Sun, our planet plows through clouds of micro-meteorites nesting along its orbital path. But as the highly esteemed authority Dr. Brian Mason of the Smithsonian Institution admits in *Meteorites*, "It is apparent that current ideas on the origin of meteorites are in a state of flux. A great deal of research work which will throw new light on the history of these complex and enigmatic bodies is at present in progress in many parts of the world. Our concepts of the origin of meteorites will probably undergo considerable changes in the next few years." One definitive statement comes from the noted Australian scientist G. J. H. McCall. Commenting on the question in *Meteorites and Their Origins*, McCall says, "We are at least certain that meteorites come to us from out of the sky—indeed they are defined as material that falls to Earth from Outer Space."

But from *where* in outer space? The location—and quantity—of these would-be meteorites has posed a headache for NASA scientists.

During the questing summer of 1972, the Space Age approached its moment of truth. Would meteorites preclude exploration of the outer planets? Such was the mid-July uncertainty faced by the control center in Pasadena, California, as it anxiously monitored progress of unmanned Pioneer 10—a $50 million question mark.

For our space probes, the point-of-no-return occurs at the moment of blast-off. Theirs is a one-way passage only. Launched the previous March from Cape Kennedy for a flyby inspection of massive Jupiter, Pioneer 10's 9-foot parasol of instruments, traveling a half-million miles a day, had slipped past the Moon in only 11 hours. Now, charging beyond the orbit of Mars (a Space Age first), it was being vectored into an unexplored wilderness known deceptively as the asteroid belt. The realm of these "little stars" between Mars and distant Jupiter is no narrow band, however, but a vast scattering of flying mountains, boulders, pebbles, orbiting the Sun in a loose doughnut shape some 175 million miles broad (nearly double the distance from Earth to Sun) and an inescapable 50 million miles thick. Despite the power of its Atlas-Centaur launch engines, the craft could not detour above or below the asteroid belt—only bore straight through. Even at Pioneer 10's record speed of 31,000 M.P.H. this segment of its Jupiter-bound journey would take 4 months.

Pasadena had to wait, hoping for the best, as banks of computers sorted precious information flowing from electronic ears of the global Deep Space Network. How densely populated is that asteroid belt out there? Is it a menace to reaching Jupiter and beyond?

For nearly two hundred years, astronomers had been studying this mysterious gap between the Red Planet and the "Shadow of the Sun": first with primitive telescopes, then with cameras, now with supersensitive electric eyes called photomultipliers; striving to calculate numbers, sizes, velocities, orbits. The task was most challenging, because scientists worked only with erratic snatches of sunlight reflected from the dark hides of the tumbling bodies. Four asteroids, they knew, bulk large enough to rate as midget planets. The largest, Ceres, boasts a girth of 480 miles, forming a jagged chunk of real estate nearly as wide as the states of Pennsylvania and Ohio combined. Next in size are Pallas, 304 miles; Vesta, 240 miles; and Juno, 120 miles. With Space Age technologies, scientists had estimated that another 3 dozen bodies measured a hundred miles or more across. The orbits of 1,831 had been plotted. Astronomers suspected that uncounted millions, ranging in size from Pikes Peak down to that of a washtub, raced unseen through space, at speeds of 45,000 M.P.H. or better.

One explanation for the enigmatic asteroid belt dates back to the discovery of Ceres and company between 1801 and 1807 by Italian and German astronomers. An Esthonian professor of physics and mathematics, Johann Sigismund Huth, of Dorpat University (now Tartu), claimed "these tiny planets were as old as all the others, and it seemed more probable . . . that the matter which formed the planets had coagulated into many small spheres in the space between Mars and Jupiter." Huth's prediction proved accurate on a modest scale, and his coagulation idea remains basically unchanged today.

Space Age scientists theorize that 4.6 billion years ago the solar nebula—an immense cloud of gas and dust—condensed into globs of primordial matter, the seeds of our solar system. Accumulating mass, these cores accreted into the Sun and its revolving planets. So gargantuan was Jupiter, however, relative to its neighbors, that its gravitational pull prevented a planet from growing between this "Shadow of the Sun" and Mars. The asteroid belt, then, is a stillborn planet. Critics of the coagulation theory dub it a "cosmic abortion."

In the beginning, according to the abortion hypothesis, the domain of little stars held a far grander population than now—perhaps 3,000 times as many bodies. Perturbed by the gravity of brawny, bad-neighbor Jupiter, the original asteroids engaged in a kind of demolition derby, smashing themselves to smithereens, powdering the space-gap with the pumice met by Pioneer 10—discharging a rubble of sandblasted rocks, boulders, and flying mountains.

Many scientists agree that this cosmic shooting gallery remains in full operation. Fragments still are ricocheting out of the asteroid belt to spiral in toward the Sun, their paths crossing the orbits first of Mars, then of Earth. With the luckless timing of a car meeting a train at a railroad crossing, wayward asteroids graze our atmosphere as fireballs, drop as meteorites, or occasionally impact with H-bomb ferocity.

No one could guess how much pea gravel and sand lay strewn in the path of Pioneer 10, but most of NASA's pre-launch calculations had been reassuring. Asteroidal material was so dispersed around the "doughnut" that hazards to Pioneer 10 would be minuscule. Even if 20 million projectiles roamed between the orbits of Mars and Jupiter, figured one mathematician, why would a single object be alone in an area measuring 180 by 180 miles. Another scientist, however, tightened chances for catastrophic collision by populating the belt with 100,000,000,000,000 asteroids big as a kitchen stove. But at these cosmic velocities, any direct encounter with a "washtub" would vaporize the fragile aluminum parasol in a puff of light, a giant step backward for mankind's space-probes of Jupiter, Saturn, Uranus, Neptune, and Pluto. Even a pebble could wreck delicate instruments, and Pioneer 10 could not dodge to escape collision. So deep into space was the craft now, that emergency instructions by radio, even at the speed of light, would take a full fifteen minutes to reach it.

Fretfully, Pasadena waited for instruments aboard the hurtling spacecraft to report back. Rimming its 9-foot, deep-dish antenna were a nest of four telescopes and a cluster of asteroid detectors. They could spot a golfball-size rock a half-mile away, a grain of sand at thirty feet. Special detector panels could sense the tiniest penetration.

In mid-July, Pioneer 10 began flashing the information that it *was* being struck—but only by micro-particles, the size of talcum powder grains. The trip through the myriad of midgets had only begun, but up to October 20, detectors aboard Pioneer 10 had counted only 83 tiny hits. The remainder of the passage remained uneventful.

In February 1973, a jubilant NASA announced that its spacecraft had safely navigated the asteroid belt. Even in the most dense region, Pioneer 10 had been struck an average of only once a day, with no particles larger than a grain of sand. The vehicle reported no damage. NASA's meticulous planning had vectored the craft away from the larger cosmic bodies by 4 million miles. "We're firmly convinced," smiled a NASA spokesman, "that the asteroid belt presents little hazard for future spacecraft going to explore the outer planets." The trail blazed in 1972–1973 through the realm of little stars has since been followed by other spacecraft. They, too, enjoyed smooth sailing; yet all have left in their wake some very bumpy questions about the space-cradle of fireballs and meteorites arriving on Planet Earth.

The assumption now feels unsteady that shooting stars, meteors, and fireballs invading our atmosphere, plus chunks of stone and steel falling as meteorites, were born in the asteroid belt. Challenging the meteor source, G. J. McCall adds that "no more than 1 percent could conceivably stem from asteroids, and the conclusion is inescapable that most very bright meteors are not of asteroidal origin."

A 1971 NASA report, *A Physical Study of Minor Planets*, further discredits the asteroid belt as headwaters for the inflow of shooting stars, fireballs, and meteorites. For an asteroid to escape the orbit of "little stars," NASA argues, the object must accelerate by an additional 5,000 M.P.H. When two space-objects collide, however, half the resulting energy is lost as heat with much material vaporized. NASA suggests it is "very improbable" that any fragment could speed up enough to flee the asteroid belt. "Thus the majority of asteroids cannot contribute to Earth's meteorite influx."

For the mother lode, some scientists look to comets. The chemical make-up of these flashy visitors to our solar system can only be guessed at: no fragments ever have been recovered for laboratory analysis. Their gaseous tails grow luminous as they near the Sun, and can reach astonishing dimensions, stretching halfway across the heavens for perhaps 50 to 100 million miles. The phenomenon, which once caused dreadful anxieties among the ancients, may endure for weeks. Unlike the tenuous tail, the comet head may be relatively tiny, from 1 to 100 miles broad. This nucleus theoretically consists of frozen gases, icy crystals of water, ammonia, and methane, imbedded with dust particles. It has been compared to a dirty snowball. But comets are such distant and infrequent travelers through our solar system that they can hardly account for the space bombardment continuously witnessed on Earth.

Wherever all this stuff is coming from, shooting stars, meteors, and "stones from the stars" do not arrive willy-nilly around the clock and across the calendar. Rather, they seem to follow a kind of timetable.

According to the so-called law of averages, the number of incoming meteorites should be evenly distributed by clock and by calendar. The hours of daylight and of darkness, as well as the months of the year, should share equally in the quota of fresh arrivals from space. Yet, records of the past century and a half indicate definite high periods and low periods. This meteorite timetable features a midafternoon peak, a late spring high, and a winter month of remarkable popularity among giant stone specimens—those space-rocks that survive the fires of entry to fall as meteorites. Their landing times on Earth seem to mock our theory of probability. Three such time-frames are inspected below.

THE 3:00 P.M. PEAK

During its yearly circuit of the Sun, our Earth makes a continuous left-hand turn. And in its orbit, our globe spins like a top. Although continents and oceans slide through alternate zones of sunshine and shadow, the side facing the Sun basks in perpetual noon, with the opposite side condemned to midnight. The front end of spaceship Earth forever rides at 6:00 A.M.—its stern at 6:00 P.M. (See Fig. 3.)

Sailing around its circular course in space at 66,000 M.P.H., our world carves out a tunnel broad as itself, engulfing all particles both great and small that are not quick enough to escape. How logical to assume that the front end of 6:00 A.M. would get splattered by meteors, like bugs hitting the windshield of a car; leaving the rear "window" relatively untouched. Yet just the opposite is true! According to records of the past 150 years, two thirds of those meteories seen to fall from the sky landed on Earth's stern. Furthermore, their arrival times were concentrated in the hours of afternoon and evening, between noon and midnight.

The greatest number fell around 3:00 P.M. During any given year, the count for this midafternoon high was 44. Twelve hours later, at 3:00 in the morning, the tally dropped off to almost zero. The second most popular time period was high noon.

Early attempts to explain this out-of-phase phenomenon centered on the velocity and direction of incoming objects. Asteroids flying toward us head-on add their speeds to Earth's, and in the atmosphere this increased velocity generates temperatures hot enough to incinerate even a sizable meteorite. Conversely, those creeping up from behind, during noon to midnight, must overtake our planet: their relatively slow speeds mean less friction, with more survivors falling as space-rocks.

New information on the trajectories of asteroids challenged the head-on collision explanation, because all bodies move in the same general direction. Solar system traffic signs read, "One Way. Left Turn Only Around the Sun."

Another less adventurous explanation for periods of increased meteorite activity suggested that more people are up and about during afternoon hours to serve as witnesses. True, an afternoon rush-hour fireball could hardly go unnoticed. And yet:

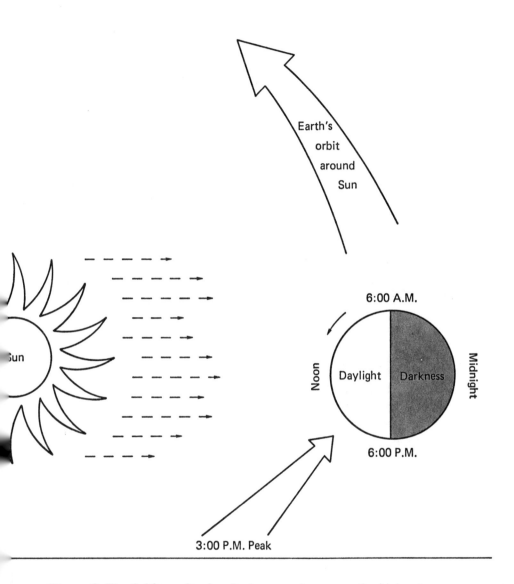

Figure 3. The 3:00 P.M. Peak During any given year, the highest number of meteorites land on Earth around three in the afternoon. (The average for this period is 44, with the count dropping mysteriously to zero at three in the morning, twelve hours later. This seems to argue that most meteorites arriving on earth are not overtaken by our planet, but rather are overtaking it.)

THE LATE SPRING HIGH

Just as puzzling as the 3:00 P.M. Peak are monthly variations in meteorite arrivals. An analysis of 581 specimens falling between 1800 and 1960 reveals that space-rocks are as seasonal as April showers, May flowers, and June weddings. Greatest numbers are counted during these spring and early summer months.

One solution to seasonal variations ventures that the frequency of incoming meteorites is constant; only the number of *observers* increases during warmer weather. But for sky-watchers, April, May, and June are no more pleasant than August and September. Another suggestion has spaceship Earth meeting periodic swarms of particles in its annual orbit of the Sun, but little evidence supports the assumption. Major meteor showers, like the Perseids and Leonids, for example, fall outside the time-frame.

The search for an explanation of these heavenly schedules becomes an exercise in frustration, making "coincidence" a most attractive con-clusion. In support of her stand, Lady Luck can point to all that space-stuff entering our atmosphere. High-altitude meteors are deceptively abundant. On a clear, moonless night the naked eye can count perhaps 10 shooting stars; an equal number burn into our envelope of air during daylight hours, only to be washed from view by sunlight. One observer, however, can watch a mere crack in the heavenly bowl. The *global* total of meteors may run as high as 100 million daily. Multiplied by 365 days in the year, the number of meteorite-candidates annually is astronomical. These schedules—the 3:00 P.M. Peak and the Late Spring High—have proven so consistent over the years, however, that chance becomes an unacceptable rationale; and we are left facing yet another cosmic mystery.

Science remains uncertain about the tally of incoming objects that do not die as shooting stars but survive as meteorites—be they micro-particles of Pioneer 10 or 60-tonners like Hoba West. Not one in a billion makes it to the ground. Survivors: perhaps 500 a year. Experts reason that nearly three-quarters of incoming meteorites disappear into the ocean. Total number arriving on terra firma during a 12-month period: only about 150. Of these, most are lost in the wilderness of mountains, deserts, icecaps, jungles. (Nobody knows how many freshly fallen mete-orites are quietly hidden from prying eyes by natives venerating them as "Messengers of the Gods.") So each year, a mere 5 or so come to the attention of scientists.

But meteorites often are deceptive in their ways, sometimes breaking their own "rules." The chart establishing that "stones from the stars" flock here during April, May, and June must be appended because it includes *all* size meteorites from pebbles to boulders. The heavyweights follow quite a different timetable. In fact, so consistent are the giant meteorites in their schedule, they might be labeled "the fireball express."

THE FEBRUARIANS

To most people, Paragould, Allende, Kirin, Sikhote-Alin, and Norton could be members of an American basketball team, mayors of minor European cities, or past winners of the Nobel Peace Prize. Mention any of the 5, however, to someone who has investigated the flights of fireballs and the arrival times of meteorites during the past 50 years, and watch the eyes snap open. It's like whispering "Everest" into the ear of a mountain climber; or "Mariannas Deep" to an oceanographer.

Paragould, Allende, Kirin, Sikhote-Alin, and Norton are big names among stone and iron space-specimens. No average-size meteorites, these, like the puny 23-pounder of Lost City. These are, according to scientific record, the largest ever seen to fall. Before disintegrating in our atmosphere, one gargantuan mass weighed an estimated 100 tons. These space-goliaths heralded their stature with slashing spears of light and reverberating sonic booms so awesome that many witnesses dropped to their knees in dread the world was coming to an end.

These champions impress us in another manner, as well: the Big Five are newcomers to Earth. All have arrived here from space since 1930. Despite the availability of 12 months in the year, four of the Big Five (like the Canadian Fireball Procession) flew in during the month of February. The lone exception was a week "late," arriving in early March.

February 17, 1930, Paragould, Arkansas: This farming community in the northeast corner of the state was jolted from sleep at 4:05 A.M. by the dazzling light and aerial detonations of a great fireball. The glare illuminated a 3-state area: Arkansas, Illinois, and Missouri, where residents of St. Louis, 180 miles north, reported an "airplane" going down in flames. A few hours after the fireball, Mr. Raymond Parkinson, a farmer living outside of Paragould, reported finding an 18-pound stone at the bottom of a freshly made pit in his horse pasture.

The main mass of the Paragould meteorite remained hidden for a month. Another farmer, Joe H. Fletcher, noticed a pit big enough to hold a cow, some 300 yards from his house by the edge of his cotton field. Fletcher had seen the fireball on February 17, as his wife knelt in fearful prayer, but made no connection between it and the new pit. Curious, he sank a pole through the mud and water, hitting something solid, "a big bump of rock." Next day, with the help of 5 other men and a team of horses, Fletcher hauled the meteorite from the pit. Originally it weighed 820 pounds, but lost several pounds to souvenir hunters and to other causes. Standing 29 inches tall, the meteorite resembles an ordinary boulder, grayish brown in color, with a surprisingly smooth surface. Blowtorch temperatures have flamed off any sharp edges it may once have possessed.

At the time, it was the largest iron or stone ever seen to fall, heftier by 100 pounds than the previous heavyweight that fell in Hungary a century ago. Also, the Paragould space-rock became the heaviest stone specimen in the world. It was purchased by Dr. H. H. Nininger for $6,200 and later was acquired by the Field Museum of Natural History in Chicago, where it remains on display today.

The Arkansas arrival reigned as champion among witnessed meteorite falls for almost twenty years. Then, 2 fresh members of the Big Five exploded against the calendar within 12 months of each other.

February 12, 1947, Eastern Siberia: A giant fireball detonated in the atmosphere some 250 miles north of Vladivostok, cascading a rain of iron onto a patch of wilderness of the Sikhote-Alin Mountains. The fireball's arrival time of 10:35 A.M. was recorded by a rural school teacher. When a blinding light flashed through the classroom window, causing children to shield their eyes, she wrote the time in the margin of a book.

Two Russian aviators, after seeing the meteor, recognized the impact point as a cluster of rust-colored craters in the uninhabited, snow-clogged woods below. The Soviet Geological Administration outfitted a small expedition for a preliminary investigation. Although a trio of scientists landed only 7 miles from the fall-site, they needed 3 days to hike the distance. Weighted down with backpacks stuffed with tents, food, instruments, and facing temperatures of 10°F below zero, the team of specialists had to push through knee-deep snows and tangled underbrush of the Siberian forest.

A fourth scientist trudged cross-country for 60 miles through the roadless wilderness. Questioning witnesses from village to village about the fireball's flight path, he used the trajectory as a direction-finder to locate his colleagues in the fastness of the Sikhote-Alin Mountains.

Investigators learned that concussions—evidently sonic booms—had shattered windows as far away as 110 miles. The fireball's track was observed as far away as 180 miles. Force of impact caused the ground to shake, with earth tremors felt 30 miles distant. To reconstruct the meteorite's in-flight behavior, Soviets interviewed over 300 witnesses. The fireball had continuously changed color, these observers reported, from red to yellow, blue to pink. In the town of Inman, some 40 miles from the trajectory, an amateur artist pictured a roiling, multi-hued trail of smoke disappearing over the wooded horizon. Today the work hangs in Moscow's Mineralogical Museum.

Scientists calculated the meteorite's trajectory by analyzing map-plots compiled from eyewitnesses and by sighting along trunks of trees knocked flat by the incoming mass. They concluded that it had originated in the asteroid belt between Mars and Jupiter, and had been in a "typical asteroid orbit" prior to entering our atmosphere at a speed of 30,000 M.P.H. The original mass was estimated at several yards wide, weighing upwards of 100 tons. Upon detonating at an altitude of 4 miles, it had separated into thousands of fragments varying from a few ounces to several tons. These segments then remained on the same course together, until impacting against what the Russians were to call "Meteorite Hill."

The impact area, atop a low, wooded hill, resembled the devastation caused by high-explosive shells. Nearly 200 craters were counted, some as wide as 90 feet, Most were ringed with uprooted trees, toppled by concussion. In shattering, the meteorite acted like a fragmentation bomb, hurling shrapnel in all directions. Trees, stripped clean of branches, stood gaunt as telephone poles. In their tops dangled broken limbs hurled aloft by the blast. Full-grown stands of birch and fir were mown down. The entire hilltop was carpeted with broken twigs and branches. The Russian Government declared the remote Sikhote-Alin area a national preserve, and began dispatching team after team of specialists there: geologists, geophysicists, surveyors, photographers, mapmakers, technicians; backed by the full resources of the National Academy of Science. Investigations continued for more than 20 years, filling volumes with technical data. The meteorite's fragments had failed to explode on contact (no evidence of heat was discovered), allowing the Russians to collect more than 30 tons of cosmic iron. The heaviest single mass weighed 1.9 tons (1,745 kg.) and remains the largest iron meteorite yet recovered from a witnessed fireball.

The Sikhote-Alin Februarian of 1947, however, rankles the mind with a trio of anomalies. First, it exploded at the abnormally low altitude of only 4 miles: 20,000 feet is well below the altitude of today's jet aircraft. Second, after detonating in the sky, why did the meteorite's thousands of fragments continue to flock together until impact? In theory, their differing weights and aerodynamic properties should have caused a wide dispersion. Atop Meteorite Hill, the Russians were baffled by the tiny "strewn field," measuring barely 1½ by ½ miles. This concentration was most unexpected. (By contrast, fragments of the 1976 Kirin City meteorite spread across 200 square miles of China.)

Even more puzzling was the timetable of the 1947 Sikhote-Alin fireball: this Februarian was followed by a second member of the Express only 53 weeks later.

February 18, 1948, Norton, Kansas: The suppertime quiet of this small farming community in the north-central part of the state, near the Nebraska border, was shattered a few minutes before 5:00 P.M. by what one witness described as "an object bright as the sun streaking through the sky"; by another as "a blinding white light in the clear sky." The fireball sighting was followed by "a terrific ground-shaking explosion thousands of feet in the heavens."

Residents believed that a military jet aircraft or a test-rocket from a Los Alamos, New Mexico, firing range had detonated in the atmosphere. Construction workers outside of Norton, veterans of World War II, were "convinced the war had started all over again." Mrs. Maud Bell, of North James Street, reported seeing a flaming object hurtling through the heavens and watching it disintegrate in a huge cloud of dust. One observer described the object as "big as a house." Reports to the highway patrol claimed that "a flaming plane had crashed."

Concussions from the mysterious aerial explosion were felt over a 35-mile area. The editor of the Oberlin *Herald* wrote that the explosion "sounded like it would cave in the roof of our shop." The Norton County Courthouse vibrated and pictures swayed on the walls. In Jennings, frame buildings shook so violently that "dust dropped out of cracks."

The Norcator postmaster heard "the almost deafening roar of the detonation," and stated that a "streak of smoke continued on, first in a straight line and then zigzagging crazily."

Two scientists hastened to Norton, Kansas, believing that the phenomenon was a meteorite. They organized search parties only to be thwarted by heavy snows blocking roads and making surrounding farmlands impassable. No evidence was discovered until the April thaw, when a 130-pound stone meteorite was found some 4 miles south of the Kansas-Nebraska line.

The main mass, however, remained hidden until July, when a Nebraska wheat farmer's tractor nearly capsized into a pit 6 feet deep. Below lay the now-famous Norton meteorite—at 2,360 pounds, the greatest quantity of cosmic stone yet recovered from the witnessed fall of a fireball. It weighed nearly 3 times more than the dethroned Paragould champion.

February 8, 1969, Pueblito de Allende: Scientists now estimate a record weight of some 4 tons of stone meteorites showered from a fragmented fireball upon the countryside surrounding this city of 10,000 in north-central Mexico shortly after Saturday midnight. Only about half that total, however, has been recovered by Mexican, American, and Canadian investigators as well as by private and commercial collectors. Thousands of specimens have been found ranging up to 50 pounds, but an equal amount probably lies scattered across the brush-covered range country or chaparral. "The recovery of this large quantity of specimen material," according to a bulletin from the Smithsonian Institution, "establishes the Allende meteorite as the largest stony meteorite shower on record."

The cascade of space-rocks was preceded by a brilliant, high-altitude fireball accompanied by tremendous detonations in the atmosphere and a violent gust of wind. Its speed calculated at between 20,000 and 40,000 miles per hour, the fireball turned night into day along a 1,000-mile track extending north into Texas and New Mexico.

The fiery mass's low-angle slanting approach before detonation resulted in perhaps the most extensive strewn field on record. It extends for about 30 miles and its total coverage may be 180 square miles—in dramatic contrast to the Siberian "rain of iron" of February 12, 1947, whose fragments were concentrated atop Meteorite Hill. Despite the spread of Allende stones, their recovery was relatively swift because they landed in such a well-populated region of the State of Chihuahua. One 7-pounder missed a house in Allende by a scant dozen feet. As the Smithsonian report points out, "The areas of maximum concentration of large specimens fortuitously and fortunately coincided with or were contiguous to areas of maximum population density." The greatest number of stones were found between the cities of Hidalgo del Parral (pop. 50,000) and Jimenez (pop. 15,000). Smithsonian scientists are making their collection of Allende stone specimens available to 37 laboratories in 13 countries.

This locale is no stranger to the fall of meteorites. "The Parral-Jimenez area," the report continues, "is undoubtedly among the richest meteorite-producing [*sic*] areas in the world. Within the relatively small area . . . three of the world's major meteorites have been recovered: the 11-ton Morito iron, the 24 tons of Chupaderos irons, and several tons of the Allende stony meteorite." Discovery of the enormous irons dates back several centuries when the region was only thinly populated. (Curiously, the Allende shower, the largest stony meteorite yet observed, landed *between* these two ancient falls.)

The Allende's twin records—for the amount of stone deposited and the extent of its strewn field—went unmatched in the world until Planet Earth greeted the Chinese contingent of the Februarians, albeit 8 days "off schedule."

March 8, 1976, Kirin City, Kirin Province, China: A flaming ball, twice the size of the full moon, shot across the sky at 3:01 P.M., Peking time, exploding into 3 sections before crashing to the ground. Workers, peasants, and soldiers by the tens of thousands watched the reddish-white meteor, rimmed with blue and trailing a whirling cloud of smoke and dust, blaze through the heavens at an altitude of about 10 miles above this northeastern province.

After interviewing more than a thousand witnesses, Chinese scientists reported that the meteorite entered our atmosphere at a speed of about 25,000 M.P.H. (12 km per second), and had been traveling through space in the same direction as Earth. During its final few seconds, the hurtling space-object exploded violently in a flash of light of sun-bright brilliance, accompanied by thunderclaps that echoed for nearly 5 minutes. Concussions cracked windows in homes and toppled merchandise from store shelves. With the blast, 2 smaller fireballs, "about the size of footballs," detached from the parent body but continued along the same trajectory to form a procession.

The main mass landed within 200 feet of a farming village, burying itself 18 feet deep amid the frozen stubble of an adjoining field. Three peasants and a group of children witnessed the impact. They first heard sounds like the noise of a supersonic jet coupled with a whine similar to that of an artillery shell. The plummeting mass struck the ground with "an appalling roar," hurling chunks of frozen earth 500 feet. Scientists are thankful it did not explode: the devastation so close to a village could have been dreadful.

Chinese investigators reported that the meteorite cracked apart on impact, and recovered from the penetration pit a large stone, blackish brown in color, measuring nearly 4 feet broad, 3 feet tall. Intact, the stone weighs 3,900 pounds (1,770 kg.), making it the largest single stone specimen in the world, exceeding even the Norton, Kansas, meteorite. Although the Allende shower of February 1969 exceeded the Kirin City fall in total tonnage, none of its fragments approach the Chinese mass in size. This new strewn field of some 200 square miles also is greater than the Mexican shower.

Officially, the giant Kirin City stone of March 8 cannot be classed as a Februarian, but its 3:01 time does qualify it for ranking in the 3:00 P.M. Peak phenomenon. In sum, then, here are dates and landing sites:

February 17, 1930	Paragould, Arkansas
February 12, 1947	Sikhote-Alin Mountains, Siberia
February 18, 1948	Norton, Kansas
February 8, 1969	Allende, Mexico
March 8, 1976	Kirin City, China

And, of course, the Canadian Fireball Procession took place on February 9, 1913.

Why February? The question can be countered only with, "Why May and June? Why 3:00 in the afternoon?" Find the answer to the Februarians, and you explain the maneuverings of the Lost City meteorite, those *Beeps* heard on a San Francisco radio station, the corkscrew trajectory of the Gulf of Mexico fireball—and perhaps also the geographic puzzle, perhaps the most surprising of all. For the mysteries are not laid to rest when these space-rocks touch down. Another matrix of enigmas is woven by their landing sites. Many irons, for example, target on the inhospitable wilderness of deserts, while stone meteorites show a penchant for congregating as close to civilization as they can.

Chapter Four

SOCIABLE STONES AND SHY IRONS

Early Scientific Scorn ☆ *Identifying Meteorites* ☆ *Caveman Space-Cache*

During the last half of the eighteenth century, European academies had been pestered by reports of stones falling from the sky. Official reaction to stories of meteorites arriving on Earth is not a proud chapter in the history of science. Experts tenaciously and repeatedly denied the very possibility that stones could drop out of the blue. Astronomers, geologists, chemists, and physicists adapted an intellectual arrogance toward a phenomenon that ran contrary to their logic and their learning—there are no rocks in the sky, therefore none can fall—dismissing eyewitness reports as unreliable and not worthy of their attention.

In defense of scientific bigotry, this was the Age of Reason, only a few stumbling steps removed from a heritage of ignorance and superstition. Accounts of sky-stones were abhorred as shameful hangovers of the past when untutored people venerated such "miracles."

Furthermore, an unformed branch of science called astro-geology decreed that for any rock to qualify as a meteorite, it should be quite unlike any terrestrial specimens. These would be the iron meteorites (actually nickel steel). Practically all irons submitted, however, had been found already sitting on the ground—not seen dropping from heaven. Stone samples added to the quandary. Although many were supposedly observed to land, they looked much like ordinary field rocks, much to the discredit of the "witnesses." This conflicting evidence, plus that vulgar belief in the miraculous origin of stones from the sky, perhaps accounts for the scientific disdain that caused some museums to junk valuable collections of meteorites.

Proof positive that chunks of stone and iron do indeed fall from above arrived via Siberia, a land where stranger things than meteorites have appeared from space.

In 1749, between the towns of Krasnoyarsk and Abakan, about 15 miles from the Yenisei River, a rural blacksmith found a roughly pie-shaped mass of iron weighing some 1,500 pounds and imbedded with nut-size, yellowish-green stones. Believing that the curious rock might contain "something better than iron," the blacksmith somehow hauled the weight into Ubeisk, where villagers recognized it as a "gift that fell from heaven." There it lay for nearly two decades, properly revered.

In 1771, the German explorer-naturalist Peter Simon Pallas visited the area and was struck by the stone's curious appearance. He arranged for its transportation to the St. Petersburg (now Leningrad) Academy of Science. The Academy dispatched samples to laboratories in other countries for appraisal. Unknown to him, Pallas had retrieved a rare type of meteorite, a stony-iron that today bears his name: Pallasite.

Hot on the trail of finding the truth about sky-stones was a young lawyer-turned-physicist, Ernst Chladni, at the University of Berlin. Ignoring popular superstition and scientific scorn alike, Chladni started his investigation from scratch by plucking his way through musty libraries and archives of Europe, digging out centuries-old accounts of "fallen masses." In his laboratory, he analyzed numerous specimens of curiously heavy rocks gathered from diverse parts of the globe, including a segment from the "Mountain of Silver," found in Argentina (and which today is lost), and a fragment of the Pallas iron.

Chladni was coming to the unorthodox but inevitable conclusion that meteorites are extraterrestrial objects. His theory was strengthened greatly by an English chemist, Edward Charles Howard, who made the link between stones and irons by finding an element common to both— the metal nickel, in a form unknown to any terrestrial rocks.

In 1794, the diligent, innovative Chladni published his findings in a short book with a long title: *Observations on a Mass of Iron Found in Siberia by Professor Pallas, and on other Masses of the like Kind, with some Conjectures respecting their Connection with certain natural Phenomena.* A storm of abuse and villification erupted from the scientific establishment, but Chladni was a scrappy fellow. Unimpressed by his opponents' prestige, he fought back, accusing them of believing that their own conceited little personalities were "the most important things in the universe." Just as boldly, many of Chladni's colleagues sided with him, mustering additional evidence that stones fall from space.

Vindication came in 1803, when even the vaunted French Academy of Science caved in—no doubt prompted by a thundering load of some 2,000 stone meteorites that landed in Normandy, practically in the Academy's lap. As Dr. H. H. Nininger jests in *Our Stone-Pelted Planet*, "It became possible for a meteorite to land in France without fear of embarrassment."

But meteorites persist in embarrassing scientists by not distributing themselves randomly across the surface of our planet. Instead, they display a strange preference for certain patches of the world's geography.

This is immediately apparent on a map, where the contrast in locations between the four stone Februarians and the single iron is startling. The heaviest iron meteorites yet found in the world have a history of being "antisocial." So was Sikhote-Alin, which buried itself in the fastness of a Siberian wilderness, blockaded by snows and miles from any village or highway. Truly a "shy iron."

By contrast, its four giant stony partners displayed a common "sociability" toward Earthlings, all landing close to human habitation. Each meteorite heralded its arrival from space with sonic thunders, as though bugling from on high, "Hey, folks! Look at me!" Then it plunged into well-populated areas for speedy recovery: The Paragould meteorite within 300 yards of a farmhouse and Kirin City not 200 feet from a village; one Allende meteorite dropping next door to a house and the Norton fireball blazing right over town. Truly "sociable stones."

In persistent disregard of our so-called law of averages, smaller specimens of stones and irons repeat the pattern of sociability and shyness. In November 30, 1954, a 3-pound meteorite crashed through the roof of a Sylacauga, Alabama, home (across the street from the Comet Drive-in Theater): The stone penetrated the living room ceiling and bounced off the console radio to strike Mrs. E. Hulitt Hodges, napping on the couch. Severely bruised on her left thigh, she became the first human on record to be hit by a falling star.

In theory, out of all the stone and iron meteorites collected about the world, a roughly equal number of each type should be seen to fall: a roughly equal number of each group should be discovered already on the ground. But in fact, the ratios are way out of balance. Sociable stones make a blazing entrance, to the gaping amazement of the audience, while shy irons sneak on stage undetected. The phenomenon is global.

Mexico: From the United States border south to Mexico City, this nation is literally loaded with giant iron meteorites, some of the world's heftiest specimens. There is no record of any witnesses. The Allende stone Februarian, however, exploded in full view of thousands of spectators: a fiesta of fireworks cascading its stonelets upon village and yard.

Russia: Since 1886, 5 stone meteorites have been collected, weighing more than 200 pounds. All were seen to fall from the sky.

India: Practically all of her more than one hundred are stones; most were observed dropping. Recent falls in 1969, 1971, 1976 were witnessed by hundreds of people.

During the past century, with the single exception of Februarian Sikhote-Alin, *all* major meteorite showers (as observed at Poltusk, Poland; Allende, Mexico; Holbrook, Arizona; etc.) have been stones, not irons. The following table illustrates this cosmic curiosity:

	Total	Falls Witnessed	Finds
STONES	1,214	766	448
IRONS	606	61	545

Most stone meteorites are *seen* to fall from the sky. Most iron meteorites are *found* squatting on the soil. Look up—see a stone! Look down—find an iron!

Why do people witness such a high percentage of stone meteorites falling from the sky—64 out of 100? Why so few already grounded? Why so few irons dropping out of the blue—only 10 out of 100? Why so many found waiting to be picked up? Scientists do offer a rationale involving different rates of weathering, and ease of identification.

The "weathering" explanation says that iron specimens, due to their high nickel content (nearly 10 percent) are as resistant as stainless steel, retaining their freshly fallen appearance for hundreds, even thousands of years. For example, the Keen Mountain, Virginia, iron supposedly fell about 1940, but laboratory analysis revealed that its arrival date was closer to 700 A.D. By contrast, stone meteorites are more apt to shatter on impact into smaller, less visible fragments. Stone meteorites also deteriorate more rapidly from rain, frost, and soil chemicals and decompose in a few years. Hence, according to this hypothesis, more iron's waiting to be picked up.

But this "decay" explanation is undermined by numerous examples of ancient stone meteorites discovered on the ground. For instance: the Potter, Nebraska, stone landed about 20,000 years ago; the Woodward, Oklahoma, stone some 14,000 years back.

The "identification" argument holds that the percentage of stone finds is low because they so often resemble ordinary field rocks: people walk right by them, but quickly recognize an iron meteorite. Yet in fact, the recognition of either type is a challenge. Every year, hopeful discoverers submit to museums of natural history and to college geology departments scores of "meteor-wrongs": cobblestones, hunks of cinders, chunks of slag, pig iron, remnants of rusted tools, and just plain ordinary rocks. Not one in a hundred is an authentic meteorite.

Neither stones nor irons are of any standard size. They may be small as a baby's thumb or elephantine, though few stone meteorites are much larger than a loaf of bread. Nor is shape an accurate guide. Some do have a blocky appearance, but usually they are irregular in outline, with little symmetry. Meteorites are not porous like cinders, nor jagged like slag: they are solid and compact, seldom with sharp edges, because the fires of entry melted and smoothed any corners.

Extremely high temperatures do leave their signature. Recent arrivals often show a charcoal-black crust, like something left in the oven too long. This thin shell is more dull than shiny. More than one innocent has cracked off the meteorite's dark skin "to see what's inside"— thus destroying much of the specimen's value to science. Other telltale marks left by the object's blazing passage through our atmosphere are curious indentations that resemble thumbprints pressed into modeling clay. Sometimes the surface is swirled like cake frosting.

Admittedly, numerous iron meteorites are more easily spotted on the ground than stones because they are not rock but metal. With time, a stone meteorite may take on the color of cracked-wheat bread, while irons superficially may appear rusty. And they are heavier: nearly three times weightier than granite. None of these clues, however, offers foolproof identification for either an iron or a stone.

The test comes in the laboratory with a grinding wheel. Scientists polish a section about the size of a postage stamp. An iron specimen will immediately reveal fresh, silvery metal, for it is nickel-steel; while a stone will show similar bright flecks. Inside color is usually gray, like cement.

Where, then, lies the explanation for the fact that without benefit of grinding wheel, and without having to witness an actual fall, so-called primitive folk of the world had no trouble telling meteorites from ordinary rock? The talent may be incredibly ancient. In the spring of 1925, workmen were clearing away topsoil in preparation for opening a new sand and gravel pit at Opava, a mining town in north-central Czechoslovakia, near the Polish border. About 3 feet below ground level, the men's shovels struck several hard objects. Unable to break the chunks apart, the laborers tossed them aside as worthless iron "bears" common to the area. An office worker retrieved a 5-pound piece as a paperweight.

Archeologists were familiar with the gravel-pit area as a prehistoric campsite dating back to the Ice Age. Alerted to the hunks of metal, these scientists excavated a total of 7, the largest weighing about 15 pounds, and identified them as almost entirely pure nickel-iron alloy—meteorites. The specimens lay arranged in a rough circle about 60 feet across. Within the circle, archeologists also uncovered the backbone of a mammoth—a shaggy, elephantlike beast today extinct. The campsite of mammoth and meteorites was dated at about 20,000 years old.

Since the 7 space-rocks did not arrange themselves in such a neat pattern, Ice Age man lugged them home—an act that raises interesting questions. How did he identify them in an Ice Age wilderness? Where did he find them? The region today is not renowned for any abundance of meteorites. In addition to the Opava finds, Czechoslovakia counts only 22, while neighboring Poland owns but 15. Their combined geographies average 1 meteorite for every 4,600 square miles. Ice Age man evidently was not tripping over heaven-stones in the vicinity of Opava.

One theory for this prehistoric cache, proposed by Lincoln LaPaz, has a primitive hunter chancing upon a parent mass, "portions of which were found by test to be easily split off from the main body of the meteorite. The fragments broken off this main mass were carried to the encampment."

This explanation is far more fragile than any iron meteorite. The alloy is one of the toughest metals known: tough enough to repel a motor-driven "diamond" saw—as an Oregon high-school teacher wearily found out in 1938. Assigned to slice off a section from the 3-pound Sam's Valley specimen, he estimated an hour's work with his equipment. The saw hardly made a dent, forcing him to apply a hand-held hacksaw, which can slice most metal. Eleven hours and 18 chewed-up blades later, he finished the job. Even diamond-tipped tools are repelled by meteorites. Needed for slicing them are high-speed saws with cutting edges of tungsten-carbide.

We may safely assume that Ice Age man had no such equipment at his disposal for cutting apart any parent mass. Furthermore, chemical analysis of the 7-in-a-circle revealed that the meteorites were not the products of any single body or any single shower. They were collected individually. Finally, because these specimens are "shy irons" and not "sociable stones," the primitive hunter of 20,000 years ago had first to scrounge among millions of ordinary rocks scattered upon the country-side, and identify these 7 as meteorites.

A fantastic accomplishment. But why go to all that trouble?

In the State of Chihuahua, in northern Mexico, linger the roofless, time-gutted walls of adobe temples and dwellings known as Casas Grandes, once the metropolis of the Montezuma Indians. According to a 1915 account by Dr. Oliver Farrington of Chicago's Field Museum, local treasure-seekers in 1867 "found, in the middle of a large room, a sort of grave with an immense block, estimated at 5,000 pounds weight, carefully wrapped, like an Egyptian mummy, in a coarse linen cloth." Actual weight of the entombed meteorite: 3,407 pounds. Size: 38 x 29 x 18 inches. Archeologists believe that the crumbling citadel of Casas Grandes flourished nearly a thousand years ago.

Indians north of the border also practiced ceremonial interment of meteorites. Dr. H. H. Nininger, in *Find a Falling Star*, relates how a curio dealer led him to a 135-pound specimen, first discovered in 1915 atop a mesa a few miles east of Camp Verde in central Arizona. They drove to an ancient ruin and came upon "a stone cyst—a little pocket in the earth, walled and covered over with flat rocks—in the corner of a decayed dwelling. The little cubicle appeared to be a typical child burial cyst, but instead of a mummy . . . respectfully wrapped in a feather cloth, a 135-pound metallic meteorite."

Specialists dated pottery associated with the little grave as between 800 and 900 years old. Dr. Nininger's analysis of the Camp Verde meteorite suggests that it came from near the giant crater further north in Arizona, adding to the perplexity of how Indians could recognize any fragments from the asteroid that impacted the desert floor. Scientists date that cosmic blast at least 25,000 years ago. The mile-wide cavity was not recognized by geologists as an authentic star-wound until about 1930.

Dr. Nininger mentions 15 other iron and stone meteorites of varying types and sizes uncovered at assorted Indian sites extending from Arizona north into Alberta, Canada; and from California east to Ohio. Obviously not all of these heavenly objects were observed to fall upon the North American continent, so sparsely peopled by the Indians. Certainly not the 3-ounce Pojoaque miniature described by Nininger as having been "found in a pottery jar in an Indian burial mound in Sante Fe County, New Mexico, in 1931." Says Nininger: "It bore evidence of having been carried in a medicine pouch, its surface indicating it had been subject to much wear against soft materials."

The specimen is a pallasite, rarest of the three meteorite categories. Only 64 of these stony-irons have been collected during the past century, and what is more: all but 11 were found in place.

Strange markings adorn the surface of the largest meteorite known to have been revered by Indians of the Southwest. This is the boulder-size Navajo iron, displayed in 2 pieces in Hall 35 at Chicago's Field Museum. The irregularly-shaped masses are dark gray color with rusty splotches. Total weight: 4,814 pounds.

In 1921, two men from Navajo, Arizona, while exploring in the desert about 13 miles from town, found the meteorite buried under rocks at the foot of a ridge. In negotiating the specimen's sale to the museum, its finders wrote, "The Navajo meteorite . . . was known to the Navajo Indians since they came to this country about 1600 (?) . . . and was covered with rocks to keep the white man or other tribes from finding it, as they thought it sacred. They call it 'Pish le gin e gin' (black iron)." The smaller of the two sections was discovered five years later, only 160 feet away, covered by dirt washed down from the ridge above. A tall rock— evidently placed by the Indians—signaled its buried location.

The surface of the larger section is most curious-looking. A groove 6 inches deep in places runs halfway around the meteorite. A museum bulletin explains that the fissure was formed ". . . not from the impact of the meteorite upon the earth, but from shock and air pressure after it reached the earth's atmosphere and prior to its fall." (For heat and air pressure to act so directionally on a fireball is most unusual, however: usually the entire surface is equally affected.)

At the time of discovery, the fissure contained many loose fragments of the meteorite, which the museum suggests resulted from impact, with additional chips flaking free with weathering. Because of the nickel-steel hardness of the Navajo meteorite, we might assume that these fragments were not chipped away by a cutting tool working on the fissure. But the surface shows definite chisel marks about 3 inches long. Their origin remains a mystery. The finders were informed by the Indians that ". . . the marks were there when they first found it, and they think the prehistoric pottery-makers cut them in."

With what?

The museum bulletin dismisses the chisel cuts, saying, "The marks

... do not appear to be anything more unusual than marks made by someone in an effort to determine the nature of the mass. It should, however, be pointed out here that the chisels used to make the marks had wider blades than those generally used now. ...."

The American Indians were a Stone Age culture with no implements for cutting nickel-steel, which defies even diamond-tipped power saws. How, therefore, did "pottery-makers" fashion a chisel some 3 inches wide, hard enough to gouge this metal alloy? Why bother? But then, why arrange meteorites in a circle? Why carry one in a medicine pouch? Why bury them like a beloved child? Is it because unlettered people recognize in space-rocks properties unsuspected by scientists? Regardless, the ability of primitives to recognize meteorites is uncanny.

Granted, most giant iron meteorites have been difficult to overlook. But who would guess they were not some ancient erratic boulder coughed up by a volcano, rolled in place by a flood, dumped by a glacier, or shoveled free by erosion? Eager to explain an unusual object in the neighborhood, superstitious minds could just as easily conjecture that some impious scamp had been turned to stone or iron by vengeful gods, rather than deduce that the weighty, wingless mass descended from thin air.

In 1576 a Spanish expedition tracked down a legend about a block of iron that had fallen from the sky. They arrived in the semidesert region of today's Gran Chaco of northern Argentina to find a group of shallow depressions in the ground, ranging in width from about 50 to 300 feet wide. Indians knew the region as *Piguem Nonraltá*, "the field of heaven," which the Spanish translated into *Campo del Cielo*. Astro-geologists recognize Campo del Cielo as a line of meteorite craters about 6,000 years old, from which they have extracted tons of iron specimens.

Indian aptitude for identifying meteorites applies to the most gigantic space-specimen yet found in United States and Canada. The famous Willamette iron, today displayed at the American Museum of Natural History, New York, measures some 10 feet long, 5 feet high, and weighs nearly 15 tons. Numerous cavities the size of washbowls penetrate its dark surface. According to legends of the Klackamas Indians of Oregon, the great mass fell from the Moon, and was regarded by them as sacred. On the darkest night before a battle, Indian braves would visit the spirit being in the forest, dipping their arrows in the holy water collected in the meteorite's cavities. Because Indian tradition mentions no great fireball heralding the huge mass's arrival on Earth, we may assume that these people used their native talent for identification.

Two centuries ago, Siberia was not exactly a seat of higher learning. Yet when a village blacksmith showed up one day with a 1,500-pound chunk of metallic rock, uneducated peasants instantly recognized it as "a gift that fell from heaven"—an origin that eluded Professor Peter Simon Pallas.

Joining in this amazing ability to recognize meteorites are the Eskimos of Greenland. From that glacier-locked island has come a host of enormous meteorites, including the largest in captivity, also on display in New York, shipped there in 1896 and 1897 by polar explorer Robert E. Peary. After much hesitation, an Eskimo guide led Peary to a coastal area south of present-day Thule, and pointed out the *Saviksue* or "Great Irons." Three made up a holy family of Wife, Tent, and Dog, expelled from heaven by Tornarsak, the Evil Spirit. Eskimo religion, however, did not interfere with Eskimo practicality. Since time immemorial they had used the Woman as their only source of iron. With rocks they unceremoniously pounded loose tiny flakes as cutting edges on their crude bone knives. The mass was surrounded by broken stone hammers.

Peary wrote in *Northward Over the Great Ice*, "The spectacle of these little fur-clad children of the ice-floes using for centuries a heaven-invented alloy (nickel steel), which is almost precisely the same in its composition as the nickel-steel armour plate with which we are protecting our battle-ships today, is to me one of the most striking in the annals of Arctic exploration."

Peary pondered how these isolated people had identified the Saviksue as meteorites, for no tradition revealed their descent from the sky. Nevertheless, the polar explorer assumed that the Eskimos had witnessed the objects' falls: ". . . else how could these rude natives have obtained any idea of their heavenly origin. . . ?"

How else, indeed? But these same arctic specimens share yet another mystery with their brethren across the planet—a geographical conundrum. For the locations of stones and irons upon the face of the globe defy logic, chance, and natural law.

Part Two
THE GEOGRAPHY
OF METEORITES

Chapter Five

DROP ZONES

World Census ☆ Favored Continents ☆ Farrington's Circle

As a map of the world instantly reveals, these space-arrivals are not scattered randomly upon the globe's 57½ million square miles of land surface: too often they lie parked in bunches.

For example, consider the Februarian space-club together on a map. The Allende shower unloaded its thousands of stones between 2 of the heaviest meteorites in the world: the 11-ton Morito giant and the 24-ton Chupaderos monster. But this is not a unique case of meteorites attracting meteorites. The Norton, Kansas, fireball landed in an area already studded with space-rocks, and not 22 miles from the Long Island stone, a previous heavyweight champion.

The map further discloses that all 5 Februarians landed in the northern hemisphere: why not below the equator? And with the whole world available as a landing site—57½ million square miles of terra firma—why did the Februarians show an obvious togetherness? Norton and Paragould stones lay hardly 550 miles apart; Sikhote-Alin and Kirin City even closer. "As we gather in more and more distributional statistics . . . ," observes G. J. H. McCall, "it becomes clear that the Earth has been preferentially, not evenly, peppered by meteorites from Space." He adds parenthetically, ". . . this was a very neglected field of meteoritics until recently."

This embarrassment of riches can be illustrated with a True-False quiz:

1. Guided by the adage "You can catch more fish with a big net than with a small net," the number of space-specimens landing on a continent or country is determined by geographic size.
2. More meteorites therefore have been found on the continent of Africa's 12 million square miles than Europe's 4 million.
3. More on South America's 7 million square miles than United States's 3.6 million.
4. More on Canada's 4 million than Mexico's 800,000.
5. More on China's 4 million square miles than on Japan's 140,000.

Each of these five statements is false! In the strange world of meteorite geography, there appears little correlation between the size of the "net" and the number of "fish" landed. The present box score reads:

China	12	Africa	125
Canada	33	South America	133
Japan	43	Europe	476
Mexico	59	United States	710

Regardless of how the global meteorite census is studied, the numbers just don't come out right. The continental United States, for example, covers about one-sixteenth the land area of the world. In theory this nation should own about one-sixteenth of the total supply of stones and irons. Rather than one-sixteenth, however, we boast nearly one-third of all known specimens on Earth.

Another shining example is the Dark Continent. Meteorites have swarmed to the most southerly tip of Africa. Two nations there make up about 7 percent of the land area, yet contribute 40 percent of the meteorite population, including world champion Hoba West.

In South America, bulky Brazil—fifth biggest country in the world—is beaten by skinny Chile, not one-tenth her size. With only 4 percent of South America's area, the Chilean ribbon claims 40 percent of her space-specimens. Chile counts more iron meteorites than all Africa. Another surprise: most of Chile's cosmic bounty is concentrated in the state of Atacama, a forbidding desert.

The geographic enigmas persist when numbers are analyzed in terms of the three basic classifications: stones, irons, stony-irons. In North America the ratio between stone and iron meteorites is roughly the same. But in Asia, space-rocks exceed space-metals by a margin of 4 to 1. In Europe, stones win, 7 to 1. In Australia, however, the tables are turned, with the percentage of irons topping stones. And in Mexico, iron meteorites enjoy a 4-to-1 ratio.

Anomalies continue with the very rare stony-irons. None have been found north of the equator in Africa, much of which is semi-arid or desert. Below the equator, Australia, similar in size and terrain, yields 8 stony-irons. There are 24 in the United States; only 4 in the Soviet Union.

Why such herd instincts among meteorites—such target-seeking talents, such pinpoint landings? Why the preference for the United States and only haughty disdain for China?

Most investigators blindly support the assumption that meteorites are touching down at random, because if our space-visitors are *not* arriving haphazardly, then ultimately we must be prepared to salute an unthinkable explanation.

So over the decades, an army of reasons has been martialed to explain the geography of meteorites, but as yet no single cause has won agreement among scientists. The answer most soothing to orthodox thinkers has discovery numbers conforming to population numbers. More people find more meteorites; higher cultures identify higher numbers.

Farrington: "Hence a map of the localities from which meteorites are known shows by far the larger part of them in civilized countries and the falls apparently the more abundant the greater the population." Elsewhere he says, ". . . the distribution of meteorites on a map of the world is almost exactly that of the Caucasian race. This seems to prove quite conclusively that the distribution of meteorites is largely dependent on the degree of civilization attained in a region."

Nininger: "If we look at the distribution of known meteorites over the earth, we find that in general it conforms to the distribution of civilized peoples."

Krinov: "This map [of the Soviet Union] demonstrates graphically the above observed dependence of the frequency of meteorite finds upon the population density and character of the area. Thus, for example, it can be seen that the largest number of meteorites is shown for the territory of the Ukrainian S.S.R., where the population density is the greatest . . . The places of meteorite finds in the Asiatic part of the U.S.S.R. are concentrated mainly along the main railroad lines, i.e., in the regions of the densest population. . . ."

Hawkins: "There is a surprising connection between the number of meteorites seen to fall [in the United States] and the number of people living in a certain area of the country."

Mason: ". . . within a specific region the recognition and recovery of meteorites depends to a large extent on the population density and cultural level."

In sum: More people find more meteorites; higher cultures identify higher numbers. But authorities have to qualify this theory when confronted by the most heavily peopled country on Earth. With a population approaching one billion, with a civilization extending back in time some

4,000 years, with every plot of ground avidly tilled, China has an inexplicable dozen meteorites. Little Japan across the bay owns 43, neighboring Russia 145, next-door India 115. On the opposite side of the globe, the United States has 710.

On no firmer ground are explanations for the abundance of stone specimens in one region, the dearth of irons in another. On the continents of Europe, Asia, and Africa, stones far outnumber metal meteorites. Rationale: since prehistoric times, people there have used up the iron, converting it into tools and weapons. But where are these hundreds of tools, these hundreds of weapons? Artifacts fashioned from meteoritic iron are museum-rare.

Like tottering old veterans on parade, one by one these solutions have stumbled, as scientists look for alternatives.

With rare imagination, Dr. McCall looks beyond Earthly factors to space influences as the reason-why, speculating that separate flocks of stony asteroids and iron asteroids intersect the Earth's track at rhythmic intervals. McCall speaks of ". . . the truth now emerging: that orbital groupings control terrestrial distributions to a considerable extent, and that successive interferences between meteorite orbits and the Earth years apart can well bring down meteorites of the same type at geographically close points."

This bold and attractive theory might explain cosmic timetables like the 3:00 P.M. Peak, seasonal spring-summer highs, and the Februarians. Except (there are always exceptions!) that the grandest of the Februarians, Sikhote-Alin, was an iron meteorite, not a stone. Furthermore, the spectacular shower of stones of February 1969, over Allende, Mexico, landed where it shouldn't have: smack in the middle of a herd of giant irons.

Chicago's Field Museum looked long and hard at the illogical clusters of meteorites upon the globe, only to mount a plaque in their hall of meteorites informing visitors:

> Meteorite falls are more numerous upon one part of the Earth's surface than upon another, though we know of no fundamental reason why this should be so. High altitude, dry climate, and populated regions have been cited as possible factors favoring concentration of meteorites.
>
> Mountains are said to exert a great gravitational force; dry air tends to keep rocks from rapid deterioration; and populated community offers better facilities for observing and finding falls of meteorites.

Yet, records show that none of these physical, climatic, or population factors have led to any appreciable concentration of meteorites, nor can there be seen a lack of concentration in contrasting areas where opposite factors are prevalent.

Perhaps there is a law governing the distribution but that law has not as yet been discovered. Nevertheless the meaning of the unusual groupings of meteorites may not be dismissed as mere coincidence.

In all the world no continent or country matches the United States in curious clusterings of stones and irons. We are the curio shop of meteorite distribution, with unexplained concentrations in the East, the Midwest, and the West. Take New Mexico, where 1 man searching 1 section of 1 county toted away no fewer than 74 stone meteorites. He is Ivan E. Wilson of the American Meteorite Laboratory, Denver, Colorado. According to the September 30, 1973, issue of *Meteoritics*, Wilson and 2 colleagues conducted a walking search of 2,300 acres in Roosevelt County. Over a period of 5 years, they picked up 85 specimens lying on the ground.

The diverse kinds of specimens and their varying degrees of weathering suggested that most members of this flock had plopped down individually, instead of hailing from heaven in a single shower. "Relying upon age and internal structural differences," reports the Denver scientist, "it has been possible to tentatively identify as many as 57 separate falls. And indications are that there may be as many as 20 more falls represented." One by one and two by two they came, targeting on 2,300 acres out of our globe's total 37 billion acres.

Although none of the Roosevelt County discoveries was especially heavy (the biggest weighed less than 30 pounds; the smallest but a fraction of an ounce) this funneling of stone meteorites on one minuscule part of the globe seems astonishing. Why should a corner of that country be so unduly favored with space-rocks?

Wilson suggests nothing out of the ordinary: "The finding of such a large number of meteorites is a unique combination of interested meteorite hunters and favorable hunting areas." During the Dust Bowl agricultural disaster of the mid-thirties, topsoil here disappeared, leaving patches of hard, bare ground. These "wind blowouts" exposed meteorites sitting on the hardpan.

In search of explanation for this local bounty, the scientist is led astray by logic. Naively, he assumes that "meteorites fall at random," and suggests that diligent hunters would find similar quantities throughout the region: ". . . it is believed a like concentration of meteorites will be found wherever the conditions are right." A highly capable investigator next is seduced into predicting, "Since 85 meteorites were found in this area, extrapolation of the figures would indicate that the meteorite population of Roosevelt County present in today's soil would be 58,000 meteorites"!

Are conditions right to the northeast in neighboring Oklahoma? During those terrible years of drought and Depression, the Oakies also watched farms disappear in dust storms; but Oklahoma today counts only some 20 meteorites. Other states of the high plains were raked by wind erosion. The Dakotas suffered from "black blizzards"; yet today North Dakota owns a mere 5 meteorites, of which only 1 was picked up following the Dust Bowl era.

Since "meteorites fall at random," why all stones and no irons from Roosevelt County? To the west, New Mexico borders the most impressive congregation of giant metal meteorites on Planet Earth.

More irons to the west; more stones to the east in America's heartland. Kansas leads the world in large stone meteorites. In addition to the 1-ton Februarian of Norton, the state claims 6 others weighing 200 pounds or better. Inexplicably the Kansas collection of some 80 meteorites (practically all stones) landed in the western third of the state.

Further east lies more "iron country." In the southern Appalachian Mountains, metal meteorites lie strewn across 6 states like beads from a necklace. Virginia, Kentucky, Tennessee, North Carolina, Georgia, and Alabama are bedecked with more space-irons than Asia, Europe, or South America.

FARRINGTON'S CIRCLE

In his *Catalogue of Meteorites of North America to January 1, 1909*, Dr. Oliver Farrington made a startling observation about the southeastern United States—a discovery that remains just as astonishing today.

At the time, Farrington—a native of Maine and graduate of Yale—was serving as curator of geology at Chicago's Field Museum, increasing its collection from less than 200 to some 700 meteorites, gathered worldwide. "The greatest massing of meteorites in the whole province of North America," he observed, "occurs in the region of the southern

Appalachians, where Kentucky, Virginia, Tennessee, North Carolina, Georgia, and Alabama adjoin." Farrington's next statement is even more astounding: "A circle with a radius of 300 miles drawn about Mt. Mitchell, North Carolina, as a center, will include nearly half of the known meteorites of North America." (At 6,648 feet, Mt. Mitchell is the highest point east of the Mississippi River.) (See Fig. 4).

Farrington analyzed and rejected available explanations for this mountain-targeting. ". . . many of the meteorites in this area might have come from a single shower . . . but the writer has made a careful study of the history of each meteorite and its geographic relation to those of similar character without finding any support for such a view." Farrington also pointed out "the finds embrace a variety of types. . . ." Neither could population density be applied as a rationale for the Appalachian concentration, "since the area is not very thickly settled."

Especially puzzling to Farrington was the large percentage of iron specimens. Prior to the 1909 census, an unexpectedly high number of metal masses had been noted in certain desert regions of the globe such as central Mexico, with scientists speculating that specimens there had been preserved by the dry air, while others in more humid regions decomposed. Farrington regarded this reasoning as untenable for the phenomenon of southeastern United States. "The climate of the region is moist, the average rainfall being 50–60 inches, so that a relatively rapid disintegration of iron meteorites might be expected."

Farrington could not hold that some "extra-gravitational force" might be attracting the space objects, nor could he accept Mt. Mitchell acting as a barricade to intercept meteorites in flight through the atmosphere. "Magnetic influences may also be suggested," proposed Farrington, reaching no conclusion.

A half-century later, Dr. Brian Mason published a state-by-state meteorite census. During the interval, the number of falls and finds for the United States had jumped from less than 200 to about 700, with the baffling high percentage of iron specimens persisting in Appalachia. Today the states Farrington cited (Kentucky, Virginia, Tennessee, North Carolina, Georgia, and Alabama) share 120 stones, irons, and stony-irons. Of these 120, no less than 85 are irons. North Carolina, home to Mt. Mitchell, owns 20 of this strange cosmic bounty. These figures become more meaningful when contrasted with the iron-count of states neighboring Appalachia.

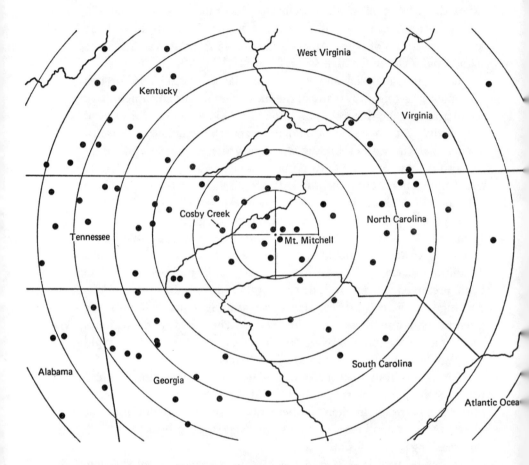

Figure 4. On Target! The states surrounding bullseye Mt. Mitchell, highest peak east of the Mississippi, have been bulleted with nearly 90 meteorites, most of them irons. The innermost circle measures 85 miles across. Arrow points to the Cosby Creek Meteorite, the heaviest iron yet found between Tennessee and Arizona.

Appalachia average	14	West Virginia	4
Ohio	5	Illinois	2
Arkansas	5	Mississippi	1
South Carolina	4	Florida	0

To account for this heavenly favoritism, an authority once speculated about a hidden body of Appalachian magnetic ore pulling in passing metal asteroids. But the region is noted for coal, not iron. The steel-making state of Alabama actually ranks lowest on the Appalachia list with "only" 8 iron meteorites. By this reasoning, Minnesota with its vast Mesabi ore deposits should own more than 1 lonely metal meteorite (2 stones, and 1 iron).

In 1837, Dr. Troost, the state geologist of Tennessee, reported discovery of an "oblong square block" weighing somewhat less than a ton. The great iron meteorite had been found 3 years before near the North Carolina border, sitting on a hillside above Cosby Creek. The unnamed discoverers tried splitting the mass with sledgehammers, "but not succeeding," quoted Farrington, "they placed it upon what is here called a 'log-heap,' where, after roasting for some time, it developed certain natural joints of which advantage was taken with cold chisels and spikes for its separation into fragments. These were put into a mountain wagon and transported 30 or 40 miles to a sort of forge and there hammered into 'gun scalps,' and other articles of more common use." Thus disappeared most of the Cosby Creek meteorite. Years later, a 5-pound fragment was seen in a Tennessee home serving as a nutcracker.

The Cosby Creek meteorite remains the largest iron ever found between Tennessee and Arizona; and its resting place lay only about 55 miles from Mt. Mitchell, central to the Appalachian gathering. Here we face twin mysteries: a wondrous accumulation of space-irons within a 6-state area, plus the largest metal specimen east of the Rockies lying in the shadow of the East's highest peak. The phenomenon left Farrington supposing ". . . some force tends to bring about their concentration here." But that force has yet to be identified.

More cosmic mysteries can be viewed along a narrow corridor stretching survey-straight down the western fringe of North America. This is "Iron Alley," where enigmas unroll like a strip-film, picturing in quick succession many of the puzzlements already observed, while offering glimpses of more surprises ahead.

Chapter Six

IRON ALLEY

North American Heavyweights ☆ *What Price Meteorites?* ☆
The Willamette Mystery

Iron Alley illustrates the unexplained herd instinct of many giant iron meteorites, plus their curious affinity for desert regions. Iron Alley portrays a return engagement of Shy Irons, whose arrival from space, unlike that of the Sociable Stones, went unreported. Central to Iron Alley is a grandiose meteorite crater, looking as if dug by some godly power with a keen eye for geometry.

The northern terminus of this cosmic corridor is spiked by the 15-ton Willamette iron, while the southern end is sprinkled with a thousand steely stars of the Xiquipilco shower. All 19 space-metals weighing over a ton on the North American continent, excluding Greenland, landed within this 2,200-mile-long, 250-mile-wide lane extending from Portland, Oregon, south to Mexico City. The remainder of the land mass—from the Canadian Rockies to the Laurentians, from Point Barrow to Key West— lies barren of giant iron meteorites.

Figure 5 shows the strange concentration of giant iron meteorites along the west coast of North America, while the accompanying table offers capsule descriptions. The remainder of this chapter elaborates on the more spectacular members of Iron Alley.

MAP NO.	NAME	STATE	TONNAGE	DATE FOUND
1.	WILLAMETTE Largest meteorite in United States and Canada.	Oregon	15 tons	1902
2.	PORT ORFORD This massive stony-iron has never been relocated.	Oregon	Approx. 11	1856
3.	GOOSE LAKE Retrieved by Dr. Nininger from boulder-strewn "Devil's Garden" near Oregon border.	California	1.3 tons	1938

MAP NO.	NAME	STATE	TONNAGE	DATE FOUND
4.	OLD WOMAN	California	3 tons	1975

Like the Willamette iron, second largest U.S. meteorite landed in a lawsuit.

| 4-A. | SQUAW TOM | California | Unknown | 1976 |

Not a member of Iron Alley in good standing.

| 5. | QUINN CANYON | Nevada | 1.5 tons | 1908 |

Oriented specimen, about three feet broad; shaped like a chocolate candy kiss, its surface partially melted. On display at Chicago's Field Museum.

| 6. | BARRINGER CRATER | Arizona | — | Prehistoric |

Uncounted tons of meteoritic metal have been carted from flanks of this 3/4-mile cavity, also known as Canyon Diablo.

| 7. | NAVAJO | Arizona | 2.2 tons | 1921 |

Hidden by Indians with rocks and boulders.

| 8. | TUCSON | Arizona | 1 ton | Pre-1850 |

Jaggedly circular and pierced by a hole. Found about 25 miles south of Tucson.

| 9. | CHISOS MOUNTAINS | Texas | 2 tons | 1915 |

Discovered in today's Big Bend National Park among its sawtooth ridges, a dozen miles from Mexican border where Rio Grande abruptly bends north to give park its name.

| 10. | COAHUILA | Coahuila | 5 tons (?) | 1837 |

Numerous iron masses have been found scattered across the salt-desert.

| 11. | CASAS GRANDES | Chihuahua | 1.7 tons | 1867 |

As previously noted, discovered in ancient Indian ruin, lovingly wrapped in linen like a human mummy.

| 12. | EL MORITO | Chihuahua | 12 tons | 1600 |

This and the three following meteorites lay grouped around the city of Hidalgo del Parral, and are studied here as the "Chihuahua Quartet."

13.	SIERRA BLANCA	Chihuahua	?	1784
14.	CHUPADEROS	Chihuahua	23 tons	1852
15.	ADARGAS	Chihuahua	3.6	1780
16.	BACUBIRITO	Sinaloa	27–50 tons	1876

Largest meteorite on North American continent.

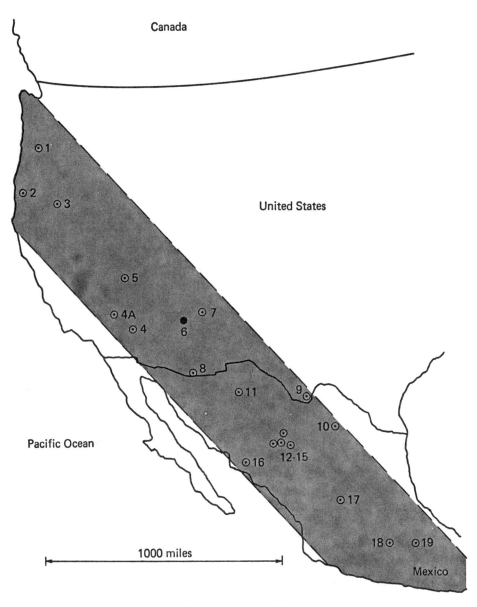

Figure 5. Iron Alley

MAP NO.	NAME	STATE	TONNAGE	DATE FOUND
17.	ZACATECAS	Mexico	1.1 tons	1792
	Years before 1792 identification by scientists, the 4½-foot long chunk of metal had lain half buried in St. Domingo Street, Zacatecas. Local accounts say it was found by an early colonist-miner.			
18.	XIQUIPILCO	Mexico	25 (?)	1776
	Meteorites by the hundreds have been taken from the vicinity of this remote village.			
19.	NATIVITAS	Tlaxcala	1.4 tons	1904

FEATURED MEMBERS OF IRON ALLEY
(from south to north)

XIQUIPILCO (Number 18): During the past 2 centuries, iron meteorites by the hundreds have been harvested some 30 miles west of Mexico City, near the village with the tongue-tingling name of Xiquipilco. Ranging in size from a few ounces up to 300 pounds, the irons have been plowed up in fields, scooped up in dry creek beds, picked up on hillsides. "Nowhere on earth have so many masses of meteoritic iron been found in one and the same neighborhood," gasped Farrington in 1904, "as in the Valley of Toluca."

Long before the first published mention of meteorite finds in 1784, settlers were shaping the malleable metal into farm tools. As early as 1825, the remote village was shipping specimens to world museums. Chladni mentions an Xiquipilco sample in the Vienna collection. In 1856 one investigator netted 69 modest specimens weighing about 90 pounds. In 1929, the ubiquitous Dr. Nininger and a colleague went on a buying spree for *areolitos*. While awaiting delivery from native entrepreneurs, Nininger picked up a 3-pounder in a field. After returning to the town square, Nininger was greeted by "men standing with baskets and bags and handfuls of meteorites. For the next couple of hours we bought meteorites until we ran out of money and the Indians were still holding up meteorites to sell . . . We bought 700 pounds. . . ."

Any estimates of the total weight of cosmic metal that avalanched around Xiquipilco are only guesses. Twenty tons seems conservative. This scattered bounty is unique, however, in that all other members of Iron Alley arrived by the ton, not the pound.

BACUBIRITO (Number 16): "It's a visitor from hell!" warned the village priest. "Not to be touched!" Such was the reaction to the discovery of North America's proudest, heaviest meteorite. Strangely shaped Bacubirito may be the second largest on the globe, bettered only by 60-ton Hoba West of South-West Africa (Namibia). Bacubirito's true weight is not available, but estimates range between 27 and 50 tons.

The *Gran Meteorito*'s original resting place lay some 700 miles north of Mexico City in the state of Sinaloa, back in the hills near the old mining town of Bacubirito. A farmer struck it with his plow in 1863. To him, a bright gash on a boulder protruding from the topsoil said silver. To the village blacksmith, it said iron. To the cleric, it said the devil. No word of the remarkable meteorite reached the outside world until 1876.

According to Farrington, Professor Henry A. Ward conducted the first scientific investigation in the area in 1902. Excavation indicated that the celestial lode had fallen thousands of years ago when the lush cornfield was bare of topsoil, its enormous weight crushing the bedrock beneath. Because there was "absolutely no trace of soil" between iron and stone, Ward assumed that "the meteorite had fallen on the bare surface of the district at a period before the vegetable soil had begun to form." By discovery time many centuries later, this loamy blanket had accumulated to a depth of 6 feet.

How, therefore, could a village priest know that the ancient mass was not of this world, and warn away his congregation? Nininger has a wry rationale: "It is possible that he shared the belief common among the local population at the time, that the mass was native silver, and that a motive other than the safe-guarding of his flock from a satanic curse prompted this advice."

For decades the great meteorite remained in place, under a protective canopy erected by the Mexican Government. In June 1959 it was moved to its present home in Culiacan, state capital of Sinaloa. There, *El Bacubirito* balances atop a white concrete block under a shade tree amid the tropical greenery of the city's Constitution Park. A plaque gives the weight as 50 tons, taken from Ward's on-the-spot estimate in 1902. Many of today's authorities, however, say 27 tons.

By any weight impressive, Bacubirito is the most unlikely-looking meteorite in the world. Not blocky, roundish, or tapered like most other specimens, Bacubirito is bent and curved: resembling a gnarled boomerang, as long as the average living room (13 feet), and a yard wide. One end is uncharacteristically squarish, as though fashioned by a giant hammer.

THE CHIHUAHUA QUARTET (Numbers 12, 13, 14, 15): What Kansas is to stony meteorites, the state of Chihuahua is to heavy-weight irons. Eight of the state's meteorites weigh over 100 pounds. Of these 8, 5 top a ton. Of that 5, 4 were found near the city of Hidalgo del Parral, within a 50-mile radius. Their combined weights exceed 30 tons, but their chemical structures differ, leaving little possibility that the members of the Chihuahua Quartet are the fragments of a single parent asteroid. Instead, each ended its own hundreds-of-millions-of-miles odyssey at practically the same spot on the third planet from the Sun. How space visitors achieve a common goal remains a most baffling mystery of drop-zones like Iron Alley.

EL MORITO (Number 12): This magnificent example of an oriented meteorite is almost a perfect cone. Three feet tall, somewhat wider at the base, it weighs 11 tons, half again as much as the 4 stone Februarians: Paragould, Norton, Allende, Kirin City. One side of El Morito is scarred by the in-flight loss of about a fourth of its original mass.

   A few parts of El Morito's history have been pieced together. Spanish Conquistador Oñates was the first European to mention the mass in 1600. For another century the "stone of iron" lay buried on a ranch or hacienda named San Gregorio. Erosion exposed El Morito to the ranch owner, who designated it as a boundary marker between San Gregorio and the neighboring spread, Villa de Allende. About 1800 he uprooted the marker and moved it closer to his home, where in 1828, a blacksmith eyed the mass as a ready source of valuable iron. Unable to cut nickel-steel, he decided to soften it in a bonfire, and set ablaze "an enormous amount of wood" fanned by 5 bellows.

   According to an observer, the blacksmith heated the meteorite to a cherry red—but then faced a 12-ton radiator. For protection against the fierce heat, he devised a shield of thick boards, and managed to chisel free a 3-pound sample, "which 3 pounds cost him $130 and they are not worth $4.00." The blacksmith's frustration yielded to pious humility, and with great effort he incised into the cooled surface the famous inscription still carried by El Morito today, and now displayed at the School of Mines in Mexico City:

Solo Dios con su poder
Este fierro destruira
Porque en el mundo no habra
Quien lo puede deshacer.

"Only God with his power
Will this iron destroy,
For in all the world there shall not be
Any one who can undo it."

Ancient Indian religious traditions about El Morito again demonstrate "uncivilized" people's amazing ability to identify iron-stones as extraterrestrial. They looked upon the object as a sacred memorial, honoring the first Indians migrating into the area from present New Mexico. Legend says that the massive weight was dragged over desert and mountain by a god disguised as a woman, bent and wrinkled with age. At his/her command the human travelers paused at the western border of the Mexican desert. One group returned north, another continued south. The deity set the meteorite upon the land as a boundary marker, a line of perpetual peace between two diverging tribes. (Legend fails to mention whether the crone-god positioned the other three members of the Chihuahua Quartet.)

SIERRA BLANCA (Number 13): Only scant information is available. Nininger's catalogue merely says, "Several masses weighing several tons." His map places the Sierra Blanca meteorite near the middle of the Chihuahua Quartet.

CHUPADEROS (Number 14): First reported by Spanish explorers in 1581, the Chupaderos twins were discovered only about 800 feet apart. Their proximity and fit-together shapes indicate a single mass split on impact. One weighs as much as the Willamette iron, the other about half that. Total weight: 45,945 pounds, or just under 23 tons.

ADARGAS (Number 15): In the mid-1800s, rumors out of Chihuahua about a huge block of metal, known for centuries to local people, prompted several scientists to investigate. The 3¼-ton Adargas meteorite was spotted "standing like a post in the earth" near the corner of a house on the Hacienda de Concepción, about a dozen miles south of the picturesque city of Valle de Allende (on today's maps: Allende). Ranchers found the meteorite between 1780 and 1784 buried in the sand. Indian lore has it falling from heaven—when, nobody knows. An unexplained date cut into the surface reads 1600. Local tradition also claims that the Archangel Malinche was secretly carrying the mass through the sky when, upon hearing the crowing of a cock, he let it fall to the ground.

Generously, let us allocate chance as the reason four of the world's largest space-irons are herded together in such a remarkably small area. But then, what explanation is left for the events of shortly after midnight on February 8 when, to the astonishment of hundreds of witnesses, a fireball deposited one of the greatest shower of stones yet observed on our planet? This Februarian hit the bulls-eye, targeting its cosmic cargo of Sociable Stones upon and around the town of Allende, home ground for the 4 Shy Irons of the Chihuahua Quartet. If coincidence is again the answer, then Lady Luck owns one beautiful bomb sight!

COAHUILA (Number 10): Years before science was alerted to this celestial bonanza, local people had carted precious hunks of "silver" into the town of Muzquiz, where a blacksmith used one as an anvil. Between 1837 and 1889, prospectors found blocks of meteoritic iron in astonishing numbers some 250 miles northeast of Allende, across the baked-out hills and salt flats of the Mapimi Desert in the neighboring state of Coahuila. One batch, known as the Butcher Irons after their discoverer, consists of 8 specimens totaling nearly 2 tons. Another investigator counted 13 within a mile of one another—one giant weighing an estimated 1¼ tons.

In reaching for explanation for this desert depository, scientists cited a fireball observed in the area in 1837. Although an Indian did carry a small meteorite into town about that date, no witnesses to any meteorite shower stepped forward. The age of the meteorites and their total numbers remain unknown.

OLD WOMAN and **SQUAW TOM** (Numbers 4, 4A): This pair's discovery in the Mojave Desert is marked by a triple legal snit and a case of mistaken identity. Late in 1975, 3 prospectors hiked into the almost inaccessible Old Woman Mountains, a national game refuge in San Bernardino County, about 175 miles east of Los Angeles, and found a metal boulder half-buried in a dry gulch. Suspecting a meteorite, they kept their strike secret for a year before finally notifying the Smithsonian Institution. Curator of Meteorites Dr. Roy S. Clarke, Jr. hurried west and identified their discovery as a rare Type II B iron meteorite: only 14 of this classification are known in the world. Its weight was later established at 6,070 pounds, a shade over 3 tons—making it the second-largest ever found north of Mexico. Due to the Old Woman's weathered appearance, Dr. Clarke believed it had landed hundreds, even thousands of years ago.

To retrieve the weight from its mountain desert hideaway, Dr. Clarke arranged an airlift by Marine helicopter to an access road near Cadiz in June 1977. It was then trucked to the Bureau of Land Management office at Riverside for public display, prior to shipment to the National Museum for laboratory analysis.

But, the prospectors argued in a federal district court, government possession of the Old Woman meteorite did not constitute ownership. In the words of the 1872 Mining Act, they had uncovered an "ore body of commercial size and value," and thus claimed that the meteorite belonged to them. The Smithsonian countered that the few tons of iron was hardly a commercial quantity; and legally no mining strike. It reinforced its case with Public Law No. 209, an act for the preservation of American antiquities and objects of scientific interest. Furthermore, the meteorite was found on federal land, within a wildlife refuge under the jurisdiction of the Department of the Interior.

Attorneys for the State of California and for San Bernardino County next joined the courtroom struggle, seeking an injunction preventing removal of the celestial mass to Washington, D.C. But as Ellis Hughes had learned in 1905, and as the federal judge now ruled anew, a meteorite belongs to the landowner—in this case, the United States Government as represented by the National Museum.

Although legally deprived of their prize, the prospectors felt entitled to a finders fee; litigation with the Smithsonian continued. As of October 1978, no sum had been agreed upon.

In dollars and cents, how much *is* a meteorite worth? There is no set price; it's a buyer's market. Sales are limited to sparse numbers of professional collectors, museums, and laboratories. They barter among themselves, and understandably try to purchase new discoveries at rock-bottom prices.

Many factors guide the purchasers—their current needs, the specimen's size, chemical structure, condition—but generally, buyers are influenced by two criteria: the meteorite's scientific value and the prestige it confers upon the owner. A freshly fallen specimen is usually valued more highly than an ancient one. Preferred are weights between 50 and 500 pounds: larger masses mean costly transport; very small ones offer insufficient material for thorough laboratory study. Shape is important: oriented meteorites are treasured. Degree of preservation is critical. Meteorites with their fusion crust intact are prized far above ones that have suffered amateurish attack by hammer, acid, or fire.

In his autobiographical *Find a Falling Star*, published in 1972, Dr. Nininger recalls his offer to sell a fragment from the Xiquipilco shower. A beautifully sculpted specimen, laboriously polished, the iron "weighed about six pounds," he writes, "and today would be worth [between] $250 and $300—or about $50 a pound. Nininger implies that the average offering cannot command such a high price, stating in *Science Digest*, "Meteorites bring $10 a pound."

In 1969, the National Research Council of Canada generously offered $1,000 for any authentic fragment from a fireball that blazed over Prince George in central British Columbia.

In 1976, Chicago's Field Museum conducted "The Great Meteorite Hunt—$100 Reward! . . . Only new finds qualify . . . No purchased specimens qualify. There is no closing date for the hunt."

At a recent flea market, a Montana college student paid a dollar for a curious-looking rock, only to sell it later for $1,200. Assuming the object the young man lugged home did not weigh over 50 pounds, the price per pound was $24.

A meteorite is worth precisely the amount someone is willing to pay.

Meanwhile, another group of prospectors, headed by Tom Sanders, age 76, reported discovery of a meteorite in the Mojave Desert's Avawatz Mountains, about 100 miles northwest of the Old Woman. This new find, nicknamed Squaw Tom, was even larger, the miners claimed. Shrewdly they refused to divulge its location until a finders fee had been placed in escrow.

Striving to strengthen their claim with political muscle, the prospectors contacted a state representative, the California governor, and a United States senator before submitting a fragment of their find to a UCLA geology professor. Despite their cagey precautions, they still came up losers. The professor's scrutiny demonstrated that meteorite identification is not for amateurs. The geologist reported that the Squaw Tom sample was of artificial material, not of meteoritic origin.

GOOSE LAKE (Number 3): In the fall of 1938, two deer hunters chanced upon this 1⅓-ton iron meteorite in the northeast corner of California, near Goose Lake on the Oregon border.

Upon news of the discovery, onto the boulder-littered mesa known as the Devil's Garden tramped the peripatetic Dr. Nininger, "through dense forest, up canyons, and over ridges." Then on horseback to the mountainous site. For transporting the cosmic cache, his party engaged a "four-ton log wagon, drawn by four heavy draft horses." Loading the washtub-size iron took hours, then trundling it from mesa-top, a day and a half: ". . . miles of almost hub-deep mud, boulders, treacherous ponds, and streams lay before us. Scouts went ahead to pick out a course, and in some cases to build a makeshift road, clearing away boulders, filling in low spots with brush, and cutting down small trees."

Like the Old Woman, the Goose Lake meteorite was shipped to the National Museum. "The finding place," reports Nininger, "was national forest land, and so this great meteorite belongs to the nation."

WILLAMETTE (Number 1): A constellation of questions unanswered shimmers around a meteorite "noble in size and wonderful in physical appearance," as described after its discovery in 1902 about 8 miles south of Portland. Weighing 31,107 pounds, the Willamette iron is the most gigantic space specimen ever found in the United States; and the strangest-looking in captivity. Ten feet long, four feet tall, the dark mass of nickel-steel resembles a huge church bell sculpted from Swiss cheese, cavities and all. Basins and channels gully its surface, a most extensive erosion by atmospheric fires.

The great Willamette iron also landed in a lawsuit—in this case pitting the rustic genius of an enterprising local farmer against a propertied Oregon corporation.

The site lies 2 miles northwest of Willamette, today part of West Linn. Following the 1902 discovery, the region was described as "a wild one, covered with primitive forest of pine and birch, little visited, and largely inaccessible." While cutting firewood during the fall, Ellis Hughes, age forty-three, noticed a curious rusty-colored boulder overgrown with brush. Striking the object with a stone, he heard it ring like a bell. From his mining experience in Australia, farmer Hughes suspected that a rich lode of metal lay below.

Assisted by a prospector companion, Hughes dug around the boulder to reveal its broad V-shape: flat side up, narrow end down. Subsequently, the two men realized they had found an enormous meteorite— but on somebody else's property. The land belonged to the Oregon Iron and Steel Corporation.

Hughes was evidently willing to let the thing sit, but his wife recognized a gold mine when she saw one. She urged her husband to move the mass of metal the three quarters of a mile through the woods and onto their farm. His companion figured the task impossible, and departed. Hughes quietly tried to buy the adjoining property. No sale. Still, the landowners had both the meteorite and the law on their side, whereupon Hughes decided to steal the iron from Oregon Iron and Steel.

A formidable project. He would have to chop a roadway through the timber, build a wagon able to support the huge weight, and devise an engine for hauling the mass. For help, he had only his 15-year-old son and an old horse. These he coupled to a remarkable engineering talent and raw strength.

Hughes spent the winter of 1902 and spring of 1903 chopping a swathe some 4,000 feet long between the meteorite and his farm. (As he later testified in court, he also cleared an extra 800 feet in the opposite direction to confuse any nosy neighbors.)

For his wagon to support the meteorite, engineer Hughes built a flatbed of 10-foot timbers with wheels sawed from tree trunks. Positioning his horse-drawn cart next to the V-shaped mass, he upended it with a system of levers and pulleys, capsized it flat-side-down on his wagon, then lashed it securely in place.

For his engine, Hughes fashioned a crude but workable capstan, resembling an oversized spool, staked to the ground and chained to a tree 100 feet from the cart in the direction of his farm. One end of a long, hand-braided wire cable he fastened to the windlass; the other end he stretched back to the loaded wagon. Hitching his horse to the capstan, he had the animal plod around in an endless circle. As the cable wound around the drum, the cart was winched forward.

Progress was pitifully slow. Some days the meteorite moved only the length of the wagon. Whenever the 100-foot cable was fully wound, Hughes had to dismantle the capstan and reassemble it another length nearer his farm. Best day's travel: 150 feet. All summer he labored. With the fall came the rain. Tree-trunk wheels sank deep into the sodden earth. Hughes laid a plank road ahead of his mired wagon.

After months of almost incessant toil a full year after its discovery, the giant iron meteorite lay in the farmyard of Mr. and Mrs. Ellis Hughes. Now they could cash in. Hughes built a display shed, announced his prize, and charged 25¢ a head. He undoubtedly intended to sell the meteorite later. From Oregon City, sightseers rode the electric streetcar some 3 miles to Willamette, then hiked another 2 to view the world's third largest meteorite, which had been discovered in their neighborhood. Soon crowds began arriving from Portland.

The November 6, 1903, edition of the Oregon City *Enterprise* printed some good news and some bad news for Hughes and his spectacle. The good news: a Smithsonian Institution representative had pronounced the dome of metal an authentic meteorite. The bad news was a rumor that the meteorite "was actually discovered on land adjoining that of the parties who now have it in possession and claim sole title thereto."

Among the more interested visitors to Hughes's farm was a sharp-eyed lawyer for the Oregon Iron and Steel Corporation who had spotted the roadway chopped through his client's property.

On November 27, Ellis Hughes was named as defendant in the "iron stone" case—without doubt the only man in the history of jurisprudence accused of swiping a 15-ton meteorite. Sadly for the rustic genius, a meteorite claim is not a matter of finders-keepers: legally part of the land, a meteorite belongs to the property owner.

Valiantly, Hughes tried to circumvent the rule by distinguishing the discovery as an item of personal property—an abandoned Indian relic—quite separate from the land itself. Two Indian witnesses from the long-disbanded Clackamas tribe testified for Hughes. Up until 30 years before, the Indians had venerated the meteorite as their *Tomanawos*, or visitor from the Moon. The holy object had belonged to Clackamas medicine men.

Unimpressed with Hughes's arguments, the court awarded possession to Oregon Iron and Steel. Hughes fought on, losing his appeal in the Oregon State Supreme Court in July 1905. For the rest of his days, Hughes harbored a bitterness about what he believed had been a gross injustice; and after his brief months of fame, lived an obscure life, dying in 1942 at age 83.

The Willamette iron remains a grand showpiece today. After being displayed at the Lewis and Clark World's Fair in Portland, a benefactor purchased it for a reported $26,000 and shipped it cross-country to the American Museum of Natural History. In 1936, the museum transferred the 15-ton attraction to its Hayden Planetarium wing where it now stands, a proud member of an internationally famous meteorite collection. A plaque gives pertinent statistics about the Willamette iron, but doesn't explain how the meteorite landed upside down.

Its bell shape identifies the Willamette iron as an oriented meteorite: rather than tumbling and spinning, it achieved stabilized flight as it penetrated our atmosphere and kept one face constantly oriented toward the Earth. Unlike an arrowhead, which travels pointed end forward, an oriented meteorite's blunt side becomes the leading face. During the final moments of its descent, the Willamette iron had somehow upended, partially burying its narrow end. Falling like the letter *A*, it landed like the letter *V*.

The in-flight behavior of an oriented meteorite is well known to Space Age scientists. In fact, this very same blunt-nose shape helped solve two of the critical problems in bringing astronauts back to Earth safely. During a return trip from the Moon, for example, a space capsule approaches our atmosphere at meteoric speed, about 25,000 miles per hour. The air friction of re-entry generates temperatures up to 100,000°F, nearly 10 times that of the Sun's surface. Space engineers had to devise a way to decelerate the spacecraft without squashing its fragile human cargo with excessive G forces, and at the same time dissipate the deadly heat capable of vaporizing the capsule itself. As *Scientific American* admitted in 1961, "The technological problems are formidable."

Following World War II, fledgling space engineers first applied the streamlining principles so successful with high-speed aircraft and the German V-2 rockets. But the V-2s had hit a top velocity of only 3,500 M.P.H. Experimental capsules sporting the familiar needle-nose were incinerated. In 1957, after more than a decade of trial and error and multimillion-dollar investment, NASA announced a breakthrough in aerodynamic design. The new space capsule sported the outline of a television picture tube.

Rather than slicing the air arrow-fashion, the ungainly blunt nose acts like a bulldozer, piling up a huge slug of air in front, braking the vehicle's cosmic speed while imposing a tolerable 9 Gs gravitational force on its passengers. The bell shape also shunts aside most of the atmospheric heat. For example, in the case of the Mercury capsule, first launched in 1961, this air brake supplied 98 percent of the deceleration and shunted aside 99 percent of the incandescent peril. To protect astronauts from the remaining 1 percent, engineers insulated the capsule's outer shell with bonded layers of resinous fiberglass—thereby accidentally discovering the principle of ablation that made re-entry feasible at last. As the fiberglass layers char, melt, and vaporize, they unload heat onto the atmosphere, by streaming a trail of fire safely *behind* the returning capsule.

With rare enthusiasm *Scientific American* could then report, "The ballistic, blunt-nosed vehicle with an ablation heat shield is one of the lightest, and certainly the simplest solution of the re-entry problem that can be achieved."

Ironically, this identical solution lay unnoticed throughout the costly, frustrating years of experimentation. As early as 1935, Dr. H. H. Nininger had proposed the idea whose time had not yet come: a blunt nose for space missiles, based on the contours of oriented meteorites. Dr. Nininger became entranced by surface features of certain pellets that had cascaded by the hundreds upon Estherville, Iowa, in 1898. "When finally the 'blunt-nose' design was adapted by the Advisory Committee on Aeronautics," Dr. Nininger recalled in 1969, "the photograph of the tested design . . . conformed almost exactly to the photograph of the Estherville meteorite. . . ."

Pellet or giant, an oriented meteorite turns its broad face to the Earth in stabilized flight, and rides out the hellish fires of atmospheric friction exactly like a space capsule. A large meteorite may ablate away tons of its outer surface before its cosmic velocity slows to 400–500 miles an hour. Then in free fall, melting stops: the surface cools and congeals, leaving a fusion crust. All the while, the interior of the fireball has remained surprisingly cool.

Picked up immediately after landing, few meteorites feel even warm to the touch, and almost never hot. Some have been found frosted over—the final exhalation of the absolute-zero cold they knew in outer space.

Science remains unaware what forces sculpt the aerodynamic perfection of oriented meteorites like the Estherville pellet and the Willamette mammoth.

Returning to the triple enigma of this nation's largest space visitor:

How did the oriented Willamette iron manage to turn over in flight and land blunt-nose up, pointed end down like an arrowhead? According to Farrington, "The mass lay buried in the ground with the point downward . . . with the apex or cone buried below. . . ."

Did the Indians upend it? Not likely. Their Tomanawos was sacred, undoubtedly never to be disturbed. There is evidence that North American tribes and other "primitive" peoples of the world regarded the ground beneath a large meteorite as just as hallowed as the celestial object itself: a heaven-Earth bond never to be broken. (Quite possibly Indians erected the adobe citadel of Casas Grandes, Mexico, around the drop-point of the 1½-ton icon, rather than transporting the celestial iron there from some distant site.)

Furthermore, Northwest Indians show no knowledge of the engineering principles and appliances Hughes used for manhandling a 15-ton weight. These tribes presumably were not megalith builders like those far to the south in Middle America and Peru.

Any engineer can argue that space capsules, despite their blunt-nose design, seldom achieve truly stabilized flight. Vehicles have a tendency to roll, wobble, and pitch, and so are equipped with a series of roll-control jets, manipulated by the astronaut. Obviously the falling Willamette iron was buffeted by air currents at denser lower altitudes, causing it to roll nose-over-tail.

This "obvious" solution, however, slams head-on into Question Number 2: Why didn't the 15-ton pointed projectile bury itself deep in the ground instead of coming to rest after penetrating the woodland soil only about three feet? Its impact was not restrained by any thick, hard stratum of stone, shale, or even clay. It landed in relatively loose soil: the same ground, when wet with rain, which swallowed the tree-trunk wheels of Hughes's wagon carrying the meteorite at rest. And on impact, this space-bullet was traveling near jet-plane velocity.

In contrast to the mysterious soft landing of the Willamette giant, a pair of much lighter stone meteorites (both Februarians) excavated sizable pits in terrain frozen solid as rock. Crashing against a snow-covered field, the 1-ton Norton, Kansas, stone burrowed nearly 9 feet down. The 2-ton Kirin City meteorite drilled a cavity almost 20 feet deep. Why should one meteorite dig its own grave, while its brethren stand monument-proud?

Let us yield to the argument that the soft landing and the upside-down position of the Willamette iron can be rationalized; that this curious behavior was the product of high-velocity air pressures, the dynamics of a free-falling iron mass, and the rebound properties of woodland soil. We still face the very existence of Iron Alley. Some of the best minds of astro-geology have attacked this puzzle, and backed away humbled.

All but 2 members of Iron Alley have been uprooted from their original landing sites, and unfortunately, few scientists have studied this major enigma within the "geography of meteorites." Explanations to date are both meager and unconvincing.

Some experts suggest a mountain-size asteroid disintegrated in our atmosphere, pounding the ground beneath its trajectory with these chunks of cosmic steel. A similar rationale—that our Earth encountered a stream of asteroids that descended all in a row—is reminiscent of attempts to account for the Canadian Fireball Procession of 1913. In either case, all members of Iron Alley should be related in chemical structure. But the Old Woman, for example, belongs to a very rare Type II B classification: the missing Port Orford giant is a scarce stony-iron or pallasite.

Another theory refers to the impacting asteroid that blasted the 600-foot-deep crater in northern Arizona: its hypothetical splash-effect hurled fragments of space-bomb hundreds of miles. For evidence, proponents cite the Goose Lake specimen, portions of whose chemistry duplicate certain iron meteorites found near Barringer Crater. But the asteroid's actual calculated flight path ran north to south over the states of Idaho and Utah to Arizona. Exploding, the hurtling mass flung fragments of itself and the underlying rock strata in a semicircle—ahead of its path and to either side. Any splash-effect also should have catapulted chunks of iron into neighboring New Mexico, but that state is bare of 1-ton meteorites.

Orthodox acounts for the reason-why behind Iron Alley yield but a single fruit: bitter-sweet riddles. No answers are yet ripe enough for plucking. Splash-effects and exploding fireballs remain the best offerings.

Dozens of other enigmas intervene, however. If the missing Port Orford giant (Number 2 on the Iron Alley diagram) is ever relocated, it will be the world's largest known stony-iron. Most curiously, it touched down not 200 miles from the 15-ton Willamette, the heaviest meteorite ever found in the United States.

The Port Orford stony-iron offers a haunting, almost sinister tale of a geologist who died too suddenly, personal journals that disappeared so strangely, and a priceless pallasite that vanished. Completely!

Chapter Seven

OREGON'S PHANTOM METEORITE

1856 Discovery ☆ *Dr. Evans' Missing Journal* ☆ *The Search Today*

LOST: 11-ton stony-iron meteorite. Irregular shape, about 4 x 5 x 6 feet. Blackish-brown surface; metallic crystals inside. Last seen on upper slope of mountain about 40 miles east of Port Orford. Finder contact Smithsonian.

Our national museum in Washington, D.C., will undoubtedly never run such a want ad. Yet it would be legitimate, because somewhere in the brawny mountains behind Port Orford, Oregon, probably within the Siskiyou National Forest, a scientific bonanza is waiting.

It's been waiting since 1856, when a government geologist, Dr. John Evans, chanced upon the massive stony-iron in the southwestern corner of the state. During the passing decades, in quest of the "lost" meteorite, probably hundreds of search parties have prowled those rock-backed heights picketed with timber, and those craggy gorges where wild rivers like the Rogue and the South Fork Coquille tumble and crash; and where Forest Service maps instruct visitors "WHAT TO DO IF LOST. Keep calm. Do not walk aimlessly. Trust your map and your compass. . . ." Prowled, and gone home denied.

Although splendid roads and trails split through Siskiyou woods and thread over its ranges, the ruggedly beautiful near-wilderness remains a challenge, if not a barrier, to treasure seekers. Rather than terrain, however, the insurmountable obstacle to relocating the metallic mass is a confusion of clues, pointing off in a dozen directions and therefore leading nowhere. This fog of misinformation and mystery has cloaked the stony-iron's whereabouts for more than 1¼ centuries, goading many scientists into discounting its very existence as mere myth. With some reason, one investigator hints darkly about a deliberate cover-up of government records showing the location of Evans' 1856 discovery.

Because of the later insinuations aimed at discrediting Dr. John

81

Evans' discovery, it is important to establish his impressively varied professional credentials: geologist, land surveyor, medical doctor, naturalist. He belonged to three elite scientific organizations: the Boston Society of Natural History, and the Philadelphia and the St. Louis academies of science. His many assignments from the government shone with high personal achievement and official commendation.

In 1849, his exploration of the "vast cemetaries of extinct animals" in the Nebraska Bad Lands so excited fellow scientists in this country and Europe that a fossil was named in his honor (*Megistrocrinus evansii*).

In 1851, he was contracted to survey the agricultural and mineral resources of the Oregon Territory west of the Cascade Mountains. While collecting samples of soils, grasses, coal, gold, lead, and such, he also computed the latitude and longitude of key benchmarks, along with a primary line of longitude—the Willamette meridian—to serve as reference for the coming public land survey. Portions of this geologic report appeared in the journal of the St. Louis Academy of Science. The work was considered of such scientific importance by the government that it was republished 57 years later.

In 1853, Dr. Evans was retained by the Territorial governor to survey a railroad route through the Rocky Mountains, and also to carry out a special geologic study. So impressed was the governor with Evans' research that he reported to Washington, D.C.: "I will close this communication by advertising, in terms of highest commendation, to the ability with which Dr. John Evans has managed the geological portion of the work, and the great contribution he has personally made in collections, and in developing the geography of the country travelled over by him." (Mountain passes in Montana, followed today by the Great Northern and the Northern Pacific railroads, testify to Evans' talents as geologist-surveyor.)

The following account is reconstructed from several sources including Evans' personal journal, Proceedings of the American Philosophical Society, investigations by Dr. Lincoln LaPaz as reported by *Topics in Meteoritics*, and by materials from the Oregon Historical Society:

On a balmy July day in 1856, Dr. Evans, 44 at the time, wades through the high grasses flowing tawny across the upper slopes of Bald Mountain, east of the Rogue River Range, some 40 miles from the frontier settlement of Port Orford. He pauses by a lonely boulder jutting chest-high from the ground. No other rocks break the surface of the sloping field. Evans notices that the outcropping measures about 4 by 5 feet. He cannot determine how deeply it's buried, but estimates its weight at 10,000 kilograms (around 11 tons). Its dark, crusted surface has evidently been subjected to great heat. With his geologic pick, Evans pries loose a corner, about the size of 2 thumbs. The jagged fragment weighs barely an ounce. Beneath the crust, the rock is flecked with silvery crystals glinting in the sunlight. He places the sample in a sack containing the day's collection of granite, quartz, gneiss, and flint. Evans checks his location on Bald Mountain in relation to the summit, to neighboring peaks, and to the Pacific coast in the westerly distance. Then he moves on.

His small party—2 French-Canadian *voyageurs*, perhaps, with horses and a couple of pack mules—picks its way north along the south and middle fork of the Coquille River, then through the lush valleys of the Umpqua and Willamette, past the settlements of Eugene City and Salem to Oregon City, the final leg of a 5-year scientific exploration of the Northwest.

The *Oregon Argus* newspaper reports on August 30, "Dr. Evans, the efficient U.S. geologist who has been engaged for several years in making a geological survey of Oregon and Washington Territories, will leave for home on the next steamer. His museum of curiosities, gathered on this coast, will make a valuable acquisition to the collection already at the capital [Smithsonian].

In his collection at Oregon City lies the fragment of the boulder from Bald Mountain. Dr. Evans is unaware that he has discovered a meteorite of incalculable scientific value. Yet when queried later by an excited colleague back East, Evans will recall the spot on Bald Mountain without difficulty.

Also with him are voluminous scientific notes, plus a personal logbook describing the many trails he's hiked—the flowers and animals he's seen, some of the soils and rocks he's sampled, the mountaintop vistas he's enjoyed during this expedition to the Territories. Who knows, perhaps someday he'll write a book about his travels.

A century later, treasure-seekers will dissect this logbook word by word.

Now in 1856, at home in Washington, D.C., he labored month after month, organizing his voluminous field notes, journals, ledgers into an official report of his 4½ years' work in the Northwest. The Oregon

Legislature petitioned Congress to publish the forthcoming document on the mineral and agricultural resources of the two territories, as compiled by "John Evans, a gentleman of high scientific attainments . . . It is desirable that the fruits of his investigation should be made known to the world."

Obviously, the reliability and competence of Dr. John Evans, scientist, are beyond question. When he said he found a stone weighing about 11 tons imbedded in the side of Bald Mountain, he found a stone weighing about 11 tons imbedded in the side of Bald Mountain.

The fragment chipped from the parent mass lay within crates of Evans' specimens shipped for analysis to the laboratory of a noted Boston chemist, Dr. Charles T. Jackson. The chemist was startled by the sample's unique appearance and its chemical composition: crystals of olivine, yielding 10.3 percent nickel, 89 percent iron, and traces of tin. Jackson suspected a meteorite. To make certain, he forwarded a particle, about 1/10 ounce, to Dr. W. K. Haidinger, international authority on meteorites, in Vienna. Haidinger verified Jackson's suspicions, reporting that the fragment was a rare stony-iron, a pallasite—similar to the one found a century earlier by its namesake, Peter Simon Pallas.

Immediately, Jackson in Boston relayed the news to Evans in Washington, inquiring about the size and location of the parent body.

Evans replied by mail: "As to the dimensions of the meteorite I cannot speak with certainty, as no measurements were made at the time. But my recollection is that four or five feet projected from the surface of the mountain, that it was about the same number of feet in width, and perhaps three or four feet in thickness. . . ." At the time of discovery, Evans could not estimate the meteorite's overall height because, ". . . it is no doubt deeply buried in the earth, as the country is . . . subject to washings from rains and melting snows in the spring, so that in a few years these causes might cover up a large portion of it. . . ."

In the fall of 1856, Dr. Jackson alerted the scientific community to the priceless discovery by reading Evans' letters at meetings of the Boston Society of Natural History. To recover the meteorite, the group recommended an expedition, led by Evans. He was enthusiastic, reassuring Jackson by letter, "There cannot be the least difficulty in my finding the meteorite. The western face of Bald Mountain, where it is situated, is, as its name indicates, bare of timber, a grassy slope, without projecting rocks in the immediate vicinity of the meteorite. The mountain is a prominent landmark, seen for a long distance on the ocean, as it is higher than any of the surrounding mountains. . . ."

During this correspondence with Jackson, Dr. Evans outlined a method for recovering the heavy mass. He suggested cutting the meteorite into sections weighing from 100–150 pounds: "The locality is about forty miles from Port Orford, in the mountains which rise almost directly from the coast, only accessible by pack mules. . . But to remove it entirely would either be impractical or involve great expense, unless indeed a river which passes the base of the mountain (Sixes River), and empties into the Pacific, should prove navigable for a raft of sufficient size for its transportation."

Summarizing the discovery for his colleagues, Dr. Jackson subsequently published a report describing how:

> . . . Doctor Evans ascended Bald Mountain, one of the Rogue River Range, which is situated from 35 to 40 miles from Port Orford, a village and port of entry on the Pacific coast, and obtained some pieces of metallic iron which he broke off from a mass projecting from the grass-covered soil on the slope of the mountain. He was not aware of its meteoritic nature until the chemical analysis was made, but the singularity of its appearance caused him to observe very closely its situation, so that when his attention was called to the subject he readily remembered the position, form, appearance, and magnitude of the mass, and manifested the most lively interest in procuring it for the Government collection in the Smithsonian Institution at Washington. . . .

Already—like the first innocent flakes of a coming polar winter—deceitful bits of information, blown by the winds of frustration, were sifting across the trail leading back to the Port Orford meteorite.

In the spring of 1860, the Boston Society of Natural History enlisted support from other scientific groups to petition Congress to finance a meteorite expedition, led by Dr. Evans. Yet Congress appeared almost hostile toward any enterprise by Evans in the Northwest. Representatives delayed reimbursing Evans for his Oregon expenses, then showed a seeming displeasure in a more painful, even mysterious manner. They ignored a directive from President Buchanan to publish Evans' massive report. The document languished in the General Land Office.

Instead of an Oregon expedition, Evans was assigned leadership of a geological survey in the mountainous Chiriqui area of Panama. During the summer of 1860, he supervised the work with his usual thoroughness.

Upon his return, his Chiriqui report was speedily approved by Congress for publication. Evans was disappointed to learn, however, that his far more valuable Oregon document, now at the Public Printer, still awaited needed funding. The trail back to Bald Mountain was almost snowed over completely now.

On April 12, 1861, civil war split the North from the South. Any congressional willingness to invest public funds in recovering some faraway meteorite vanished in the smoke of military and political turmoil.

On the following day, Dr. John Evans, the one man capable of recovering the Port Orford meteorite through private subscription, was dead at age 49. History does not reveal the cause. One investigator of the Port Orford meteorite conjectured that Evans' death resulted from "hardships suffered while serving as geologist on the Chiriqui Surveying Expedition." But Evans, on his return from Panama, had expressed delight with the healthful climate of the Chiriqui highlands, speaking of plans to move there permanently.

Evans' trail now was erased irrevocably, as his massive report—still at the public printer—inexplicably disappeared. It would not be seen again. As for his many journals, field notes, and observations, compiled during 5 years in Oregon, only tantalizing scraps of his personal logbook, plus his few letters to Dr. Jackson, survive. Rather than guide, these fragments will perplex anyone seeking to recover the 11-ton Port Orford meteorite, waiting somewhere on the slope of Bald Mountain.

Today, more than 1¼ centuries later, the Port Orford meteorite remains lost. If recovered, it will be a giant among pygmies. Only 2 stony-irons in the world weigh more than a ton: Bitburg, Germany—1.5 tons; Huckitta, Central Australia—1.4 tons. At 11 tons, the Port Orford's value to science is beyond calculation.

Why can't it be found?

The Siskiyou National Forest is no city park, but more than a million acres of mountains, ravines, timber, and brush, containing uncounted crannies for hiding a meteorite. Its position remains inscrutable because all available clues wander aimlessly across the map. The treasure-seeker must first identify the Rogue River Range and Bald Mountain. Diligently, many astute investigators have tried and failed.

One of the most persistent searches for Evans' mission journal and the Port Orford meteorite was conducted for many years, beginning in 1934, by the noted meteorite expert Dr. Lincoln LaPaz. LaPaz complained, ". . . no one from the most venerable pioneer to such modern map experts as the President of the Oregon Historical Society . . . has been able to identify anywhere in Oregon a mountain bearing this name."

Over the years, names on maps have changed. Some hopeful treasure-hunters insist that the Bald Mountain of Evans' time is Iron Mountain today; others claim Bray Mountain. A 1973 Forest Service map shows a Bald Mountain about 9 miles southeast of Port Orford, and a Bald Knob some 20 miles east. Evans specifically mentions a distance of 30–40 miles from Port Orford. But was he speaking in the straight lines of the surveyor, or the winding trails followed by pack mules? LaPaz believed trail miles, and concentrated his search on the close-in Bald Mountain, without success.

Should Evans' peak ever be identified, the meteorite may still remain hidden, covered with rock and gravel washed down by rains and melting snows. Perhaps it has become dislodged, tumbling off the mountains and into some inaccessible ravine.

In his search, LaPaz met an obstacle new since Evans' day: his grassy slopes of 1856 have become brambled barricades of brush, as high as 8 feet in many places. LaPaz related how "progress in any manner through the intricate network of brush was laborious and slow, since neither crawling beneath the interwoven branches nor 'swimming' atop them across the prickly leaf-sea . . . accelerated progress. In many spots, visibility was restricted to less than ten feet." One of his party tunneled into the brambles at 9:00 A.M. and surfaced at 5:00 P.M. Distance traveled: a mere 2½ miles.

Another scientist, dogging the same elusive clues as LaPaz, tramped a different section of the Siskiyou National Forest. In the summer of 1939, Dr. E. P. Henderson, then Associate Curator of Meteorites at the Smithsonian, used as his guide Dr. Evans' handwritten logbook. Henderson matched the trail outline with modern topographical maps, concentrating his search some 55 miles from Port Orford, between Johnson Mountain and Bingham Mountain. Henderson identified this as the area in which Dr. Evans twice mentioned Bald Mountain in his July 21 logbook entry. Henderson, feeling he had walked where Evans had walked, and camped where Evans had camped on that date, failed to find the meteorite, like all others before him.

Because of the many failures, readers of the *Oregonian* subsequently were electrified to read: "Famed Orford Meteor—Long Mystery of Coast—Believed Discovered." According to an interview with a Mr.

Robert (Bob) Harrison, the meteorite lay on his mining location claim, between 30 and 40 miles southeast of Port Orford, "and about 12 miles from Powers and about 5 miles from the Forest Service road at China Flats." Harrison described the large rock, buried in the ground, as weighing from 15 to 20 tons. "The stone is so hard and so tough that it is almost impossible to break off pieces of it," he said. According to the *Oregonian* story, Harrison said that he had succeeded, and described the sample as being "webbed with silvery lines with brilliant spots." Harrison said that a specimen had been identified as meteoritic.

Did Bob Harrison discover Oregon's phantom meteorite? Perhaps; but that *Oregonian* story is now more than 40 years cold. It ran July 21, 1938.

Professional scientists like LaPaz and Henderson have been replaced by a dedicated troup of amateurs called the Port Orford Meteorite Association (POMA), organized by the editor of *Oregon Outdoors*, "Digger" Costelloe. Although POMA has no membership list, no bylaws, no dues, no assets, and no liabilities, these amateur treasure-hunters are mountain-wise and persistent. Almost every year, they resume the great search, fully equipped with geologic picks, cameras, campers, compass, maps, barbecue grills, and a gung-ho spirit. Their goal: to find the Port Orford meteorite, with a little fun along the way.

POMA enthusiasm, however, has brought no success. When Digger invited me in 1976 to "come out and have a go at it," he admitted: "I can take you to Bray Mountain, Bray Ridge, Tim Creek, and all the rest of the places. I can stand you in Dr. Evans' very footsteps—but I'll be damned if I can show you the Port Orford meteorite! I spent many years looking for it—I don't know where it is, but I know a hell of a lot of places where it isn't."

Despite their amateur status, POMA members are as familiar with Evans' personal logbook as any scientists, knowing full well it may be a forgery. So, one more mystery clogs the trail to the missing meteorite: the logbook's real author is unknown. Handwriting samples from it, according to an FBI analysis, cannot be identified. "The majority of the handwriting of Evans' log . . . was not written by him," the report concludes. Did Evans dictate his daily commentary to some member of his small party? Certainly not to a French-Canadian *voyageur*. As Henderson observes, "The log is an unbound, handwritten document [sixty-eight typewritten pages] measuring 12½ x 8 inches. Its appearance indicates that someone was transcribing a previous document and had difficulty deciphering portions of it. Surely an experienced explorer, such as Dr. Evans, would not carry unbound paper into the field to keep notes on." They'd scatter with the first breeze.

Who was the transcriber? Did that person delete mention of the meteorite's precise location? The question is not irresponsible.

Dr. Lincoln LaPaz invested years searching through documents entombed at the General Land Office, the Public Printer, and Library of Congress, hoping to find the original logbook and Evans' missing geologic report. Their loss, he implies, may not be accidental, but rather the result of a "deliberate interference at a high level." Details of Evans' 1856 expedition to Oregon have yet to be recovered.

A sister mystery slinks around his previous exploration of the Northwest. In 1851, he pinned down the latitude and longitude of key benchmarks for the coming survey of public lands. He also fixed the position of the Willamette meridian as a reference line that passes within 4 miles of Oregon City, where he later took departure with his "museum of curiosities" in 1856. Evans was intimately familiar with the terrain near Oregon City. Across the Willamette River—hardly 2 miles away—lay the huge 15-ton Willamette iron.

In addition to the world's largest stony-iron meteorite, did Dr. Evans also discover the biggest iron between Mexico and Greenland? Again, we will perhaps never know. All his notes of the 1851 expedition—like those following—were "lost" while in transit back to Washington, D.C.

The many enigmas and false clues cloaking the location of the Port Orford meteorite understandably have induced some scientists to question its existence. Several authorities insinuate that the specimen received by Dr. Jackson in Boston came not from any parent mass, but from the hand of a friendly Indian who appreciated Dr. Evans' interest in strange rocks.

Others humble the Port Orford stony-iron as mere legend. In 1964, Dr. Henderson wrote, "A critical analysis of the meager and conflicting data could well lead one to consider the possibility that the large mass commonly referred to as the 'Port Orford meteorite' is a myth."

Twelve years later, in response to my query, a Smithsonian curator answered, "The evidence that the Port Orford meteorite ever really existed is poor in my view. I have no information indicating that it has been found, and I doubt seriously that it ever will be." Obviously, our national museum is not about to run any want ad encouraging a search for a "mythical" meteorite. Still, this official pessimism should not be faulted; not after 150 years of fruitless quests down false trails.

Because of Dr. Evans' professional credentials, however, we can feel assured that somewhere east of Port Orford in the "Rogue River Range," the world's largest stony-iron meteorite awaits the caress of its fortunate finder.

Two final enigmas about the Port Orford involve the geography of meteorites: with some 57½ million square miles of land area as a target, why should it mark the most easterly border of Iron Alley? And why did it elect a spot not 200 miles from the Willamette iron?

Nor is this the continental United States' only major drop-zone: yet a third exists in the very heartland of America, some 500 miles from the eastern edge of Iron Alley.

Chapter Eight

THE KANSAS COLLECTION

Stony Meteorites Galore ☆ Rare Stony-Irons ☆ The
Explanation "Unthinkable"

For a state whose motto urges its citizens, *Ad astra per aspera* ("To the stars despite adversity"), the biography of Kansas is strangely incomplete.

Granted, the Sunflower State's lusty past and prosperous present are well chronicled. Who has not thrilled to Wild West movies of Dodge City and Boot Hill, Kansas? Who has not marveled at brawny, suntanned Kansas, first in wheat harvests? Who has not cheered the industrial, political, footballing Kansas? But who ever heard of Kansas' unique distinction as Meteorite Metropolis of the World—an honor that remains little known except to a batch of baffled scientists?

Reversing the state's stellar slogan, space-rocks by the score have seemingly adapted the motto, "Down to Kansas, and damn the fires of entry!"

For number, size, and species of meteorites, the Sunflower State blooms alone in the world. Her present count of 80 is bettered in the United States only by triple-her-size Texas, and by half-again-as-big New Mexico—but only if Roosevelt County's freakish finds are catalogued individually, not grouped as a shower or two.

In the global sweepstakes, Kansas has unearthed more meteorites than Sweden, Norway, Finland, Denmark, Greenland, Poland, and Portugal combined. The Kansas census equals the total of the British Isles, the Netherlands, Belgium, Switzerland, Austria, Bulgaria, Romania, and Greece. Four-hundred-mile-long Kansas has yielded more space-specimens than have been found between Morocco and India—a distance of some 5,000 miles—and claims more meteorites than the 7 South American nations of Venezuela, Colombia, Bolivia, Peru, Paraguay, Brazil, and Uruguay.

Moreover, her stones outnumber her irons by the wild ratio of 15 to 1. And no ordinary stones, these; for Kansas has harvested world champions. Until the arrival of the Kirin City giant in 1976, a pair of Kansas sky-

prizes had ruled supreme for 85 years. Her half-ton Long Island heavy-weight held title from 1891 to 1948, when the huge Norton meteorite boomed earthward, landing barely 25 miles away. Right up there on the scales with the Long Island and Norton champions are the 700-pound Hugoton stone, 600-pound Morland, and 200-pounders Admire and Farmington. Considering that an average stone meteorite weighs less than 25 pounds, these are enormous; yet more than two dozen Kansas offerings beat the average. Gather all the stone meteorites yet found in 32 of the 48 states, and you still fall short of the Kansas extravaganza.

Unlike the "sociable stones" so often seen fireballing down from the sky elsewhere in the world, most of the Kansas cascade of fallen stars are like the "shy irons" found already resting on the ground. Nearly a half-century ago, Nininger remarked that in a dozen agricultural states, 25 out of 30 stone meteorites were seen to fall, while in Kansas the ratio was only 6 out of 19.

The series of remarkable discoveries began in the latter part of the nineteenth century, when the state was settled only sparsely and farmers began busting the buffalo-grass sod with horse-drawn plows. The first authenticated find occurred in 1874 near the village of Waconda, in Mitchell County. The dull, iron-black 100-pounder was spotted on the grassy bank of a ravine and later identified by a college professor as a stone meteorite containing 5.3 percent nickel and of recent arrival. The Waconda announcement stirred up little interest, however.

The next identified space-rock was a two-way maverick: an iron (not a stone) found in the western (not eastern) part of Kansas. Farmer Quincy Baldwin, living a mile outside Tonganoxie, found a cobblestone-size rock weighing 26 pounds. Its heft and rusty color made him suspect a vein of iron ore below. To test the metal he fashioned a fishhook, while envisioning riches. But there was no deposit of ore, and Baldwin's "mine" was abandoned before it opened.

On June 25, 1890, another Kansas rarity showed up in the northeastern part twenty-five miles from Farmington. Shortly after the lunch hour, following a hearty meal at the Clifton Hotel, F. F. Woodruff was standing outside on the porch when he heard a thundering in the cloudless sky, almost directly overhead. Rolling booms drowned out the din of a passing steam locomotive of the Missouri Pacific Railroad. After 2 or 3 minutes, the noise faded.

Fifty miles from Farmington, John Yates of Grant Township felt a shock "like a hundred-pound cannon shaking the house and rattling windows." One hundred and thirty miles distant, John D. Randolph of Cedar Junction sighted an "aerolite" and reported, "It was a ball of fire as large as a table. It had a trail like a comet and it wobbled like a kite."

On the outskirts of Farmington, J. H. January lay on the ground repairing his wagon when a noise made his horse skittish. Looking up, January saw an object smash into the ground, sending chunks of dirt flying barn-high. He and 4 neighbors spent several hours digging the mass from the gumbo soil. The Farmington stone meteorite was a rough slab some 19 inches long and weighing 197 pounds. Alerted by the fireball and booming noises, hundreds of people arrived from miles around seeking souvenirs. Mr. January obliged with a sledgehammer, bashing part of the meteorite into handy-sized keepsakes.

Most improbable of all the discoveries was made piece by piece by a Kansas bride with a keen eye for iron-stones. In 1885, Mary Kimberly and her husband, Frank, homesteaded in Brehnam Township, Kiowa County. Before the arrival of farmers this land in southwestern Kansas had been cattle country—table-flat grassland, free of rock and tree. While breaking the sod with horse and plow, Frank Kimberly's share began whanging against heavy black rocks hidden beneath the turf. To his astonishment and ire, Mary insisted on saving the iron-stones, some weighing 100 pounds and more. She piled them in a shack behind the house. Recalling a grade-school museum visit, she recognized the telltale signs of meteorites: the blackish crust, the rippled surface, the greenish crystals inside laced with silvery metal.

For 5 years her "rock pile" mounted while she sent off letter after letter to geologists, asking appraisal of her finds. Finally, in 1890, a college professor responded, recognizing the heap as extremely rare stony-irons. Buying a half-ton of them on the spot, he dispersed them to waiting museums and laboratories throughout the world.

With their heavenly reward, the Kimberlys bought the adjoining property and began plowing up more stony-irons; Mrs. Kimberly gathering smaller fragments in a bushel basket. In all, their "meteorite farm" delivered about 1½ tons.

Still the crop of specimens came up. In 1923, collaborating with the Denver Museum of Natural History, Dr. Nininger began digging up 867 more pounds. Returning in 1933, he excavated a "buffalo wallow" that had been a watering hole for cattle, uncovering 1,200 pounds. He reasoned that several centuries ago, an asteroid had burst on impact, blasting out the 55-foot crater and scattering fragments for acres around. In 1947, an amateur collector toting a new-fangled metal detector in a wheelbarrow struck a 740-pounder on the old Kimberly meteorite farm. Another investigator hit even bigger with a half-tonner. Few records were kept, and no one knows how many people have wandered out with armloads of meteorite fragments. This area of Kiowa County has yielded at least four tons of "rare" stony-irons.

But stony-irons still remain in short supply, with some 64 known about the world. The largest 2 whole specimens weigh about 1½ tons each: one is in Germany; the second in Australia. The Kansas crop, therefore, can be classed as a world champion, exceeded only by the 11-ton Port Orford stony-iron, unseen since its discovery in 1856.

While Mary Kimberly was harvesting iron-stones, Kansas made another contribution to the record books. Until 1891, the world's largest stone had been an 1866 Russian contribution weighing one third of a ton. Then Kansas took over with the half-ton Long Island, which held its title until 1948 when the Norton Februarian landed not 25 miles away.

By 1900, while most states lay barren of known meteorites, Kansas already counted 13: 1 iron; 2 stony-irons; 10 stones. In 1906, in west central Kansas, a Ness County surveyor watched meteorites being collected by "the wagon load." To pay for his $2-an-acre land, a farmer hired a boy to follow the plow, picking up meteorites, which then were selling for $1 an ounce.

Farrington's 1909 catalogue mentions fifteen Kansas meteorites, stating, "Those of the western part of the State are all stones." In 1933, Nininger observed, "Of stony meteorites whose falls were not witnessed, one-third of all North American finds have been in Kansas; and about a sixth of the finds of this type of the entire world have been within her borders." By that year, the stone count had jumped to 25. Between the productive years of 1890 and 1948, while most other states were counting their meteorites by two's and three's, Kansas was ticking them off by the dozens: 65 in all, of which 59 were stone specimens. Today Kansas's cosmic bonanza totals 80, still practically all stone specimens.

Why so many meteorites in the Sunflower State? That question has been tantalizing scientists decade after decade. Why so few seen to fall? And why, unlike Appalachia to the east and Iron Alley to the west, does Kansas excel in stone specimens?

As a world authority on meteorites also personally familiar with Kansas, Dr. Nininger deserves attention when he says in *Our Stone-Pelted Planet* that ". . . it has at times been suggested that some subtle force operates to concentrate incoming bodies in this area. That any such force exists not only seems . . . unthinkable, but on a careful examination of the facts proves entirely unnecessary to a satisfactory explanation of this disproportionate share which is credited to the state." As his own explanation, Nininger pointed to the Kansas soil, "comparatively free from terrestrial rocks," making any meteorites conspicuous.

Because Kansas leads in wheat production, frequently-tilled soil might explain *how* her stones are revealed, but not *why* they are there. As contrast, Iowa lies in the heart of the midwestern corn belt; yet with 95 percent of her land under cultivation, she has gleaned but a single meteorite in this century. Illinois has plowed up only 6 in her entire history.

Nininger next explained that Kansas inhabitants have become "meteorite minded." This "interest factor" stimulates search and recognition of meteorites. "So far as is known, one part of the Earth's surface is as likely to receive meteoritic falls as another; but that our knowledge of such falls depends upon the presence of human beings intelligent enough to report them. Again we may rest assured that the nature of the soil and the degree to which it is cultivated are important factors in the discovery of meteorites. It is the author's belief that the education of the public as to the importance and value of meteorites will prove the largest factor in the delivery of meteorites in the future."

His 1933 prediction about future discoveries in Montana illustrates the quandary faced by scientists attempting to rationalize drop-zones: "The great State of Montana with its immense area has certainly been the 'landing field' of more meteorites than has Kansas, and yet only one has been found, and that an iron, while Kansas has twenty-three to her credit."

According to the most recent available catalogue (1966), Montana now has 3, against Kansas' 80. Explanation is lacking for this continuing favoritism.

A clue? We recall that Mt. Mitchell seems to serve as the focus for the widespread stone falls that form "Farrington's Circle" of Appalachia. My own investigation began with a simple question: What's so special about the Sunflower State? In all the world, what natural or man-made feature is truly unique to Kansas?

Would the geographic position of Kansas, smack in the "center" of the 48 states, somehow influence the number of meteorites arriving from outer space? That famous point is monumented with a stone cairn and plaque near the town of Lebanon in north central Kansas. But any correlation is hard to come by, because the spot is more a tourist attraction than scientifically determined point—more whimsy than geodesy, as the United States Coast and Geodetic Survey will confirm. With some amusement, engineers determined the country's "geographic center" by dangling a cardboard cutout map from a wire. The balance point happens to be in Smith County, Kansas.

And if meteorites are somehow "targeting" on this fancied location, what about Rugby, North Dakota? By the same "scientific" technique, a spot nearby is recognized as the geographic center of the entire North American continent. How does that state's meteorite population shape up? North Dakota: 6; Kansas: 80.

Then, in a surveyor's manual, I found that Kansas does enjoy a certain international distinction, and on a January afternoon inspected the little-known site atop a lonely hilltop pasture known as Meades Ranch, in Western Kansas.

From the winter-browned grass poked a low dome of concrete about 3 feet across, set with a saucerlike bronze plaque stating that it was the property of the U.S. Coast and Geodetic Survey. The science of geodesy concerns itself with determining the size and shape of our globe: information critical to surveyors, mapmakers, mariners, pilots, and astronauts.

Cardboard cutouts are not for Meades Ranch. No point on Earth has been fixed in terms of latitude and longitude with more impeccable precision by surveyor's transit, by orbiting satellites, and by computer than the benchmark at Meades Ranch. It is the master control point for all surveying and mapping operations in North and South America.

Since my visit to western Kansas, I have roamed the world through maps and manuals, seeking possible connections between the "divine science" of geodesy and this planet's unexplained concentration of space-specimens. Growing evidence suggests that an ancient and "unthinkable" marriage secretly links meteorites and mankind: a marriage performed on the altar of geodesy.

Perhaps some unknown "law" still eludes scientists despite 2 centuries of meteorite investigations around the globe, despite computerized insights into celestial mechanics. Unless that elusive law comes to light, and as veteran explanations topple one by one, there may stand before us soon only Nininger's "unthinkable."

Native "superstition" surrounds many members of the puzzling Iron Alley corridor. The Casas Grandes meteorite and the Navajo iron were both revered by natives, as was the Willamette giant or Moon Stone. El Morito was set in place by a "god" as a boundary marker. Adargas was dropped by an "angel." Barringer Crater was blasted out by a "fire god." According to tradition, the El Morito boundary-meteorite could actually speak to the Indians—perhaps in the same language that a modern geodetic benchmark "talks" to surveyors? Have "primitive" people been allowed knowledge about heaven-stones denied our supposedly sophisticated world?

Chapter Nine

THE HERMIT KINGS
Wilderness-Seeking Giants: The Silver Camel ☆ The Greenland Family ☆ "Field of the Sky" ☆ The Great Chinguetti Iron

Alone and aloof—beyond or at the frontiers of civilization—stand 7 of the globe's meteorite monarchs. They are big, each 17 tons or more. One is reportedly bigger than all known specimens combined. While lesser meteorites throng together by the score—peasants at a marketplace—these most majestic of space-visitors stand apart from the earthly mob and from one another, in wilderness areas remote from major population centers. Five elected for their drop-sites hostile deserts or semi-arid domains uninviting to human settlement. One imperial isolationist chose for its throne a wasteland of ice. All save 2 have defied removal to museums. Continent by continent, these are the Hermit Kings. (See Fig. 6.)

AUSTRALIA

Like so much of the "Desert Continent," the state of Western Australia is flat and dry. Its southeastern corner is especially forbidding, shunned even by the aborigines. Here stretches for hundreds of miles the Nullabor Plain, a name derived from the Latin words *nul* and *arbor* for "no trees." Underlying the Nullabor Plain, a porous limestone bed hundreds of feet thick creates an unquenchable sieve, sucking the surface dry after a year's occasional shower. Only tortured shapes of saltbush and clumps of leathery grasses dot the gray, stony, floor-flat expanse.

The only towns are spaced along the water-bearing Trans-Australian Railway. Here the railroad track reaches horizon to horizon, unbending for 300 miles. During construction of the lines in 1915, water was imported by camel train. To the north: nothing but stone and sand for a thousand miles. South: more desert, ending at the Indian Ocean. Four hundred miles east shimmers the Outback town of Coober Pedy, world-famous for its opals, where mining families live in subterranean grottoes

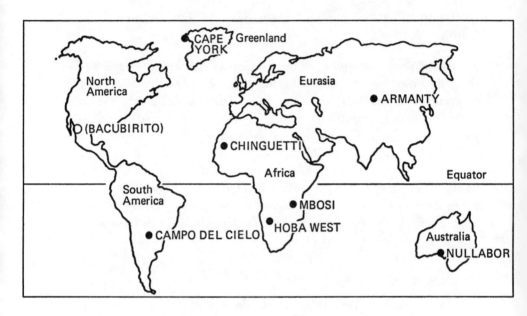

Figure 6. The Hermit Kings

to escape the broiling heat. Even religion has gone underground with the Catacomb Church.

An equal distance west of the Nullabor Plain where our first Hermit King was discovered, the sedate city of Kalgoorlie flourishes on gold shoveled by the ton from the desert floor and on water pumped through a 300-mile pipeline. Before gold was first found in 1893, Western Australia was a land of "sun, sand, sin, sorrow, and sore eyes" and fewer than 30,000 people for an area about one third the size of United States.

In March 1965, two shirt-sleeved geologists were surveying the rock strata and landforms of the Nullabor Plain. Scanning the barren flatness with telescopic instruments for something to serve as a bearing-marker, they spotted a dark object in the distance. Investigating, they found Australia's largest meteorite, a huge iron lying in two sections about 600 feet apart. The specimen had evidently split just before impact.

A photograph suggests the larger portion to be about 7 feet long and 3½ feet high. It is an oriented meteorite, shaped like an ice cream cone, deeply pitted all over. Its weight is given as 12 tons: the smaller segment weighs 5 tons. Despite its massiveness, the Nullabor meteorite—or Mundrabilla, as catalogs list it—made a soft landing, penetrating the limestone desert floor a mere half-inch. Around it, a thick layer of rusty shale indicates that the space-iron arrived eons ago. One report says, "probably thousands of years, and possibly hundreds of thousands of years." At 17 tons, the Nullabor meteorite is nearly 3 times the heft of Victoria, the previous Australian heavyweight discovered in 1854. Nullabor's smaller part was sectioned for distribution to research laboratories around the world: the newly-crowned King Nullabor the Greater is displayed at the Western Australian Museum.

EURASIA

According to Chinese sources, what may be the world's second largest museum-meteorite now rests amid the mud-walled dwellings and frame buildings of dusty Urumchi, remote capital of Sinkiang Province. Resembling a 10-foot-long baked potato taller than a man, the lumpy mass of nickel-steel may weigh as much as 30 tons—a precise figure is unavailable. (The great Greenland meteorite at New York's Hayden Planetarium bends the scale at 34 tons.) Chinese call their discovery the Armanty meteorite.

Its landing site is vaguely given as 300 miles from Urumchi in the Gobi Desert, hardly the garden spot of Asia. In Mongolian, *Gobi* means "waterless place." Fifteen hundred miles long and some 500 broad, this arid wasteland is encircled by mountains that block moisture-laden breezes from the Indian and Pacific oceans. Stones and barren hills characterize the discovery area.

Sinkiang Province is about double the size of Texas, with half the population. Most people there live in isolated cities like Urumchi or around the occasional oasis. The land beyond is virtually deserted, with perhaps a dozen inhabitants per square mile—most of them passing through. These are the nomadic Kazakhs, driving sheep, horses, and camels from one infrequent watering hole to the next.

Unlike Australia's Nullabor Plain and Africa's Sahara, which know only heat and no real cold, the Gobi offers fierce extremes of both. July may broil with temperatures of 110°F and January freeze at 30° below zero. Then the Kazakhs say, "Ice is our bed and snow our blanket." From them Chinese scientists ultimately heard about the Armanty meteorite. Nomads related that it had arrived from heaven about 200 years ago. They called it the Silver Camel.

Some camel! To move it, Chinese engineers built a 24-wheel trailer hauled by a heavy-duty diesel tractor. The Silver Camel arrived at Urumchi in October 1965, where it rests as an outdoor attraction on a circular stone platform.

GREENLAND

Of all the Hermit Kings, none reigns in isolation more splendid, nor in a fastness more defiant, amid a climate more vicious, than a meteorite known only to the Eskimos of northwestern Greenland. Although his Woman, Dog, and Tent have been kidnapped to the Hayden Planetarium, the Man himself continues to elude all.

Copenhagen's Mineralogical Museum proudly displays a 20-ton dome of pitted metal extracted from that glacial desert in 1964 and crowned the Man: obviously a pretender to the Arctic throne, because the Man must outweigh this princeling by at least another 20 tons. Meanwhile, the Eskimos' grandest heaven-stone remains hidden from the eyes of the *kabloona* (white man), who began prying into their secret source of iron in 1818.

While sailing up the northern Greenland coast, British polar explorer Captain John Ross ventured into a gulf some 100 miles below today's Thule, naming it Melville Bay. Near Cape York he met a lonely band of Eskimos untouched by civilization. They marveled at his *oomiaksoah* (big ship), inquiring if he was a visitor from the Sun or the Moon.

Ross was astonished to see their harpoon heads and small, crude knives tipped with thin, sharp flakes of iron, much like flattened nails. At first, Ross assumed that the metal came from some shipwreck cast upon these desolate shores. Vaguely the natives mentioned *Sowallick* (Iron Mountain) on the northern coast of Melville Bay, perhaps 25 miles distant. Protruding from the mountain, they said, was a dark rock about 4 feet long, imbedded with pebble-size pieces of iron. By laboriously beating the rock with a tough green stone, they broke off sections, then hammered the nodules flat.

Ross suspected a meteorite, but bad weather prevented further exploration. He settled for a rock-sample from the mountain—a gift of the Eskimos.

Was the Eskimo language misunderstood, or did the natives mislead Ross with deliberate half-truths? Their sample was not metal, only one of their rock-hammers. Later, their source of iron proved to be no immovable mountain but a mere 3-ton meteorite—the Woman, part of the family of Man. Whether through lack of communication or otherwise, the first step in safeguarding Man's location had succeeded. For the next 76 years, expedition after expedition from Britain, Denmark, and the United States visited Cape York, and generation after generation of Eskimos refused to divulge the Iron Mountain's position.

In May 1894, part of the secret was finally unveiled to *Pearyaksoah*, as Eskimos called the indomitable Lieutenant Robert E. Peary, who later credited himself as the first man to attain the North Pole. "Great Peary" was convinced that the so-called Mountain was a giant meteorite. His magnificent exploit of crossing the 600-mile-wide Greenland icecap by snowshoe and dogsled impressed the natives: his gifts of metal implements seduced them. Knowing that "their sole and ever-besieging enemies were the demons Hunger and Starvation," Peary bribed a hunter named Panikpah with the offer of a rifle. Panikpah agreed to lead Pearyaksoah to the Iron Mountain, a week-long trip by dog team. After only 2 days, however, Panikpah quit, complaining about the weather. His replacement, a guide named Tellakoteah, soon had Peary perilously offshore, skipping from ice cake to ice cake. Pearyaksoah got soaked to the waist. Regaining the land, "my driver takes me [on] a long wild goose chase," Peary later recalled, "to the front of the big glacier and after various wanderings points to a big snowdrift as the site of the stone. I tell him he is a fraud and return to the sledge." Tellakoteah also threatened to abandon Peary, but was bluffed into continuing the trek to Cape York—and the Iron Mountain's secret place.

After 11 hazardous days on the trail, Peary arrived at a small peninsula butting into the ice of Melville Bay. "Standing here," Peary wrote in *Northward Over the Great Ice*, "the eye roams southward, over the broken ice-masses of Glacier Bay, the favourite haunt of the polar bear; eastward, across the glacier itself, to the ebony faces of the Black Twins, two beetling ice-capped cliffs, which frown down upon the glacier. . . ."

The man who had crossed the ice cap regarded the climate here as fierce. "In winter this region is the desolation of Arctic desolations, constantly harassed by biting winds, and every rock deep buried beneath the snow, swept in by these winds throughout the long dark night, from the broad expanse of Melville Bay, and piled in drifts, which in many places are hundreds of feet deep. Even in summer, only the directly southward-facing slopes of the mountains are free from snow for a few weeks, while in the valleys and on the northward slopes the drifts remain eternally. . . ."

Digging through the snow, Tallakoteah uncovered the "long-saught [*sic*] object," and "at 5:30 Sunday morning, May 27, 1894, the brown mass rudely awakened from its winter's sleep, found for the first time in its cycles of existence the eyes of a white man gazing upon it." Revealed in the snow pit was a chest-high, sack-shaped *saviksue* ("great iron") weighing three tons.

Tallakoteah revealed to Peary more than three-quarters of a century of explorers had ever heard. Three *saviksue* existed in the area; this was the Woman. Nearby, somewhere under the Snow, lay the Dog. About six miles away on a Melville Bay island stood the Tent, largest of the trio. Eskimo legend claimed that they had fallen from heaven, expelled by the Evil Spirit, Tornarsuk.

Only the Woman supplied metal for Eskimo harpoons and cutting edges. Raising a small boulder in two hands, Tallakoteah demonstrated the technique of incessantly pounding away at one spot to loosen a metal flake. The Woman was softer and more malleable than either Dog or Tent. But now that *Kabloona* knives were available from explorers and whalers, she remained unmolested. Already planning recovery of the valuable *saviksue*, Peary photographed the Woman, erected a cairn marker, and map-sketched the area.

He returned the following year to retrieve the 2,500-pound Woman and 1,100-pound Dog. Eskimos cautioned him against moving the heaven-stones, however, recalling the near-tragedy of a previous attempt. After years of chipping, the Woman had shrunk to about half her size, and her head had broken off. Eskimos from the Etah settlement had seen it as a handy source of iron at home, and loaded it on a sled. While crossing frozen Melville Bay, they had barely escaped with their lives when the ice suddenly broke and sled and dogs were dragged into the depths by the weight. Since then only fragments had been removed from the Woman: Tornarsuk was not a devil to be trifled with. Peary ignored the warning by rolling and dragging both Woman and Dog to the shore and ferrying them on ice cakes to his waiting ship. The Tent proved a bit more difficult.

In 1896, Peary again arrived at Cape York—now with the *Hope*, a steamer with ice-breaker bow, which he berthed at a natural rock pier. The 10-foot-tall, 7-foot-broad Tent would tax all his engineering talents just to move it 300 feet to the water, let alone load it aboard ship. He resorted to hydraulic jacks, capable of lifting 30 tons each, to upend the mass, laboriously rolling it to the pier. With each turn, the metal "monster" would crush rocks beneath that sent out "a stream of sparks" as they "dissolved into dust and smoke." A half-inch-thick iron chain was flattened like putty.

"There were many incidents of the work to suggest the supernatural . . . ," admitted Peary. He called the meteorite "demoniac." During the 10-day ordeal, savage snowstorms howled at the sailor and Eskimo workers. Evidently Tornarsuk was getting angry. The Iron Mountain towered above the men, "standing out black and uncompromising," wrote Peary. "While everything else was buried in the snow, the 'Saviksoah' was unaffected. The great flakes vanished as they touched it, and the effect was very impressive. It was as if the giant were saying, 'I am apart from all this, I am heaven-born, and still carry in my heart some of the warmth of those long-gone days before I was hurled upon this frozen desert.' "

Peary was denied his cosmic prize this time. Fearing that September ice, already forming in Prince Regents Bay, would imprison the *Hope* all winter, he fled Cape York. Three years later he was back, doing battle with Tornarsuk and his mighty weapon, the weather.

In August 1897 as the *Hope* approached Meteorite Island, Peary realized "the Arctic winter had already set in." All Melville Bay was "an indescribable labyrinth of icebergs." While his vessel lay moored 18 feet off the rocky pier awaiting the enormous meteorite, a gale dumped a foot of snow on ship and shore. During the next 5 days, as sailors and Eskimos toiled, the weather was "one constant succession of fog and driving snow."

For loading the huge weight, Peary bridged the gap between deck and dock with massive timbers weighing 3 tons each, reinforced by double sections of railroad track. One end of the structure rested on the *Hope*; the other lay beneath the grudging meteorite up on blocks. While awaiting high tide, Peary draped the American flag over the huge iron, and his baby daughter "dashed a little bottle of wine against it and named it 'Ahnighito.' Then the jacks . . . pushed it steadily forward to the edge of the pier."

Tornarsuk unleashed another weapon. Out in the bay, a huge iceberg collapsed into the sea, sending waves crashing ashore. "On these the *Hope* rolled and danced like a cork, jerking viciously at her moorings." Peary realized that "if one of the lines parted, the great timbers, with one end still resting upon the *Hope*'s heaving deck, would act as irresistible levers to pry the blocks from under the meteorite and let it topple over the edge of the pier into the water." But the waves subsided, leaving Peary still shaken 48 hours later.

Slowly jacks and winches inched Ahnighito onto the creaking bridge toward the *Hope*'s deck—but with Eskimos no longer aboard. They feared the "mountainous weight of the 'heaven stone' would crush the *oomiaksoah* [ship], and they preferred to say farewell from the shore." Numerous sailors shared the Eskimo superstition that the meteorite would "take the ship to the bottom." Finally the brown monster was safely aboard the *Hope*, and all hands congratulated themselves on a job well done. At that moment—miraculously—a patch of blue sky appeared overhead. "It was as if the demon of the 'Saviksoah,' " gloated Peary, "had fought a losing fight, accepted the result, and yielded gracefully."

Tornarsuk had only begun to fight. A fresh gale herded gigantic icebergs toward shore, threatening to smash the *Hope* against the anvil pier or at least trap her there for the next 9 months in a vise of ice. A single ship-wide channel led between 2 towering bergs and the relatively open sea beyond. Tornarsuk's trap was closing. The iron-prowed *Hope* must ram her way to freedom. But would the captive monster, still on deck, ignore the jarring collision; or would it lurch from its supports, plunging through the ship's bottom, sinking all?

Cautiously at first, the *Hope* tried to force the twin barriers apart, only to come to a dead stop. She backed off and tried again at full speed. Blocks of ice flew through the air. The meteorite quivered. And again. The meteorite trembled. With her propeller racing and black smoke pouring from her funnel, and with the *Hope* herself "pulsating like a human heart, . . . inch by inch we squeezed between the frozen blue rocks on each side. . . ." Peary turned to watch the two bergs swing slowly together again. The exit from Meteorite Island was closed for the winter, but he had escaped the trap with his "celestial prisoner," which was "now throbbing amidships, as it had never throbbed since that cataclysmic day when it hummed through the burning air, and shook land and sea with the frightful fury of its impact."

Tornarsuk was no quitter. Other icebergs forced the *Hope* perilously close to shore and onto a reef. Danger would come with the falling tide, leaving the *Hope* tilting on her rocky perch. With unwanted tons of steel topside, "no earthly power could keep her from capsizing." Fortunately the tide was flooding, and the *Hope* floated free of the grasping reef.

The open sea of Baffin Bay offered no haven. While men labored to lower the metal mass into the ship's hold, Tornarsuk struck with a series of Arctic gales, and "the furious wind howled . . . as if the demon of the 'Saviksoah' were shrieking at us." Sailors feared that the "demoniac iron had broken loose and was smashing a way for itself through the ship's side."

Now getting low on fuel, the steamer *Hope* set a course south through Davis Strait for Nova Scotia. The metal mass so confused her compass, however, that she had to grope her way along the Labrador coast and entered Sydney burning her last ton of coal.

On October 2, 1897, the *Hope* arrived at the Brooklyn Navy Yard with her cosmic cargo. Was Tornarsuk hovering nearby? One report claims that the great meteorite nearly ended its journey on the harbor bottom when the crane hoisting the mass collapsed. Eighty horses were needed to pull Ahnighito through New York streets to the American Museum of Natural History. There it reigns today at the Hayden Planetarium, in company with the Woman and the Dog, as well as with other noted specimens like the Willamette iron, largest ever found in the United States.

The exact weight of Peary's "Star Stone of the North" remained unknown until 1956. Estimates had ranged up to 90 tons. Then the Toledo Scale Company devised a special rig and watched the pointer stop at 68,085 pounds, a trifle over 34 tons, making it the world's heaviest meteorite ever transported for public display.

Ahnighito's sale price of $40,000 enabled Peary to finance his North Pole expedition. The physical agonies he later endured and the controversy still shadowing his claim as discoverer of the Pole hint that Tornarsuk, and not *Pearyaksoah*, was the winner in this cosmic contest after all. Furthermore, although Peary did carry off the Tent, the Woman, and the Dog, he failed to track down that stellar Hermit King, the Man.

Because of the trio of landing sites on the edge of Melville Bay, and the more curious fact that neither the ground beneath nor the meteorites themselves were scarred by the impact (all three enjoyed a "soft landing"), many scientists theorize about a parent asteroid disintegrating near Cape York. They then turn to Peary's hypothesis: that originally the meteorites had "descended upon the surface of the then much-expanded ice-sheet, and upon its recession had gradually settled to the positions in which they were found." (The Greenland ice cap is but a remnant of a continental glacier that once covered almost all Canada too.) Or, Peary supposed, "one of the enormous snow-drifts which form along this coast even in ordinary winters might have received the meteorites and cushioned their fall completely . . . allowing their gradual descent and final deposition upon the underlying rocks."

A remarkable asteroid, for unknown to Peary, it deposited 2 additional meteorites, today displayed at Copenhagen's Mineralogical Museum. In 1926, a Danish expedition returned from Cape York with a rough pyramid of iron weighing 3½ tons; then in 1964, they brought back the rounded 20-ton mass they dubbed the Man. Is this the "Hermit King" of the Arctic, or does a sixth still more majestic meteorite still reign undetected in Cape York's frozen wasteland?

One investigator claims that Tallakoteah informed Peary about a *saviksoah* even larger than the 34-ton Tent that lay far inland. The Eskimo guide had visited it as a boy with his father, but either couldn't remember or refused to divulge the huge object's location. To him this great iron was the Man.

SOUTH AMERICA

Like a small-scale Iron Alley, a string of meteorites and miniature impact craters some 600 miles northwest of Buenos Aires, in the semi-arid Gran Chaco region, form the royal courtyard of Argentina's Hermit King. Like Oregon's Port Orford, this South American space-specimen has also been lost—for 200 years.

A mysterious lode of iron—or even silver—enticed earlier expeditions into the subtropical wasteland. In 1526, lured by a legend of silvery riches, Venetian explorer Sebastian Cabot entered the broad mouth of a muddy river flowing into the South Atlantic at today's Buenos Aires. With high hopes, he named it *Rio de la Plata*, "River of Silver," but found no treasure in its tributaries. The legend lived on.

In 1774, a series of Spanish expeditions began searching for a rumored silver mine. Multinamed Don Bartolome Francisco de Maguna returned with a few pounds of metal—only iron, but of extraordinary quality. The lode became known as *Mesa de Fierro*, "table of iron." Estimated weight: 25 tons.

The last man to lay eyes on the hidden trove was naval lieutenant Don Miguel Rubin de Celis. Ordered by the provincial governor to determine once and for all the type and quantity of metal in the Gran Chaco, de Celis's 1783 investigation proved most thorough. From the protruding mass he hacked off 2 hunks weighing about 5 and 25 pounds—at the expense of 2 axes and 85 chisels! To examine the soil beneath, he blasted the earth with gunpowder. To find its underground extension, he tilted the great weight with levers. But in upending such a heavy load, he may have capsized it into an adjoining pit, effectively burying the monarch iron from future investigators. It has never been seen since.

Today the location and weight of the South American Hermit King are both unknown, despite decades of search by Argentine and American scientists. This part of the thinly-populated Gran Chaco semidesert is brambled with thorn bushes, reminiscent of the brushy barricades on Bald Mountain that hide the Port Orford stony-iron.

In a land nearly as flat as an airport runway, a score of unique shallow depressions pock the ground, forming a corridor some 45 miles long and about 2 miles wide. The largest depression is about 300 feet across and 15 deep. Within this "little Iron Alley," scientists wielding metal detectors have shoveled up hundreds of iron meteorites weighing from a few ounces up to 4 tons. "In some parts of this area it is impossible to search with a mine detector for more than a few seconds," wrote the noted meteorite authority Dr. William A. Cassidy in 1965, "without finding a new meteorite. A sampling of much less than 1 percent of this central area yielded more than 500 specimens." In 1969, Dr. Cassidy reported the discovery of an 18-ton space iron, buried 15 feet deep. Was this the legendary table of iron the Spaniards sought? Not likely: according to Cassidy, "It would have some artificial markings on it where they cut off samples with chisels, and this new find does not."

The orthodox explanation for this clustering of craters and meteorites is familiar: about 6,000 years ago an asteroid detonating in the atmosphere punctured the ground with larger fragments, cascading its smaller remains as meteorites. Demonstrating the tenacious accuracy of primitive legend, Indians called the region *Piguem Nonraltá*—in Spanish, *Campo del Cielo*; in English, "Field of the Sky."

AFRICA

A triumvirate of Hermit Kings shares sovereignty over the Dark Continent. While only sparse information is available for Mbosi and Hoba West, the third meteorite represents one of the most baffling space-mysteries on Earth. After all, how do you "lose" something the size of a 12-story apartment building and weighing upwards of 1 million tons?

Mbosi: This Hermit King of incomplete biography remains where it landed near the village of Mbosi in southwest Tanzania. The site lies in east Africa's Great Rift Valley between 2 of the world's deepest lakes, Tanganyika and Malawi (formerly Nyasa). The terrain is rolling grassland, the climate hot and dry, with only some 2 inches of rain between May and October. Population is scanty. Discovery date: either 1930 or 1931. Estimated weight: between 25 and 27 tons.

Hoba West: On the opposite side of the continent, where the Kalahari Desert fringes into the skimpy grasses of South-West Africa (Namibia), practically no rain falls for 9 months of the year. Here reclines the world's largest extant meteorite, named for the landowner at the time of its discovery in 1920. This elephantine brick-shaped mass measures some 10 feet long and 8 feet wide. Originally it may have weighed up to 100 tons; but centuries of weathering have skinned it to about 60 tons. With a nickel content of some 20,000 pounds, the Hoba

West iron was threatened by smelters until the Government of South-West Africa wisely declared it a national monument.

Despite its bulk, this giant penetrated its limestone drop-site by only about 3 feet. Like other great space-irons—Ahnighito, Nullabor, Bacubirito, Willamette—its terminal velocity was strangely low, permitting a nice, safe landing.

Chinguetti: If all the iron, stone, and stony-iron meteorites listed in world catalogues were heaped together, the pile would hardly reach knee-high to this Hermit King Supreme. The given dimensions of the Great Chinguetti Iron—120 feet tall, 300 feet long—make it the biggest asteroid known to have survived impact with Earth. This million-tonner was first seen by a European in 1916, but known to native Bedouins of the western Sahara Desert 50 years before, who venerated it as "the stone that fell from heaven."

A mystery to match its size still shadows the most majestic of Hermit Kings. Its location in Mauritania is known within 50 miles. The area around the Chinguetti oasis has been searched by meteorite expeditions, probed by oil prospectors, surveyed by geologists, and mapped by aerial cameras. And yet no trace of this steely goliath has been found. Undoubtedly there are nomadic Bedouins aware of the massive meteorite's hiding place, but they have remained silent all these years. The last man to talk died immediately after, of poisoning. Whether suicide or murder, who knows?

As with Dr. John Evans, finder of the now-lost Port Orford, it is important to establish the credentials for the only non-Moslem to whom the existence of this space-visitor was revealed. Neither scientist nor explorer, Captain M. Ripert was an officer in the French Army. At the time, Mauritania was a colony of French West Africa and Captain Ripert served in the Adrar Camel Corps. Later he became Chief Administrator of the Colonies, and upon retirement, a planter in Cameroon, French Equatorial Africa.

The possibility of a hoax is minimal because Captain Ripert treated the discovery so casually. In a 1931 letter to the Governor General of French West Africa, Ripert recalled that his find "did not seem to be of any particular importance." The following year he wrote to the director of Le Verrier Observatory at Marseille, "I could not realize at the time all the interest that this meteorite might arouse. I spoke of it incidentally, when I was leaving Mauritania, to one of my friends, M. H. Hubert, a

Doctor of Science at Dakar, who did not seem to attach any great importance to its existence."

Nor is it likely that Captain Ripert, a desert-wise soldier, erred in his observations about the object's unprecedented enormity. His letter to Marseille continued, "I was able to state that the meteorite formed a sort of cliff...the surface appearance of the large mass was in no way comparable to that of the blackish polished surface of the rocks ... found on the 'reg' [rocky plain] and on the sandstone plateaus of the Adrar [region]."

As final evidence that the Great Chinguetti Iron is real, Captain Ripert returned from the discovery site with a 6-pound specimen of authentic meteorite.

One evening Captain Ripert had overheard his Bedouin soldiers discussing a strange mass of metal located out in the desert. Curious, he later approached his chief camel driver, Sidi Ahmed Zein, who denied all, aghast that a European knew about "the stone that fell from heaven." But his commanding officer persisted. Finally, the Bedouin soldier admitted the object's existence, reluctantly agreeing to guide Ripert into the desert for a quick inspection.

But Captain Ripert could bring no compass! No notebook, no pencil, no tape measure, no map; nothing to help him to record the holy object's location, size, and appearance. To further restrict Ripert's chances of ever retracing their path, Sidi Ahmed insisted that they travel at night, with darkness blotting out landmarks.

After riding for 10 hours on their camels, and as dawn turned the sea of rolling sands to pink, the officer noticed ahead a sloping "dune" some 300 feet long and 120 feet high. He thought it just another sandhill until the sun rose. Before him there flashed a vertical cliff of jagged, silvery metal, sand-blasted to a glistening finish. Galloping to the top of the sandy ramp that buried all but the object's broad face, Ripert noticed an exposed corner bristling with spikes and spears. Leaping from his camel, Ripert tried twisting one free. It would bend, but not break. He picked up a "rock" for a hammer, but Sidi Ahmed was near panic. "We must leave, Sir! Now!"

Clutching his rock, Ripert dashed back to his camel. They rode away. Returning to Chinguetti, the army officer wrote down all he could recall about his furtive night ride into the Sahara. Like the great meteorite, his notes were never seen again. And for betraying the meteorite's position,

Sidi Ahmed paid the ultimate penalty: he died soon after, "apparently of poison."

Ripert remained baffled that his discovery was not relocated. "I am astonished," he stated in his 1932 correspondence with the Marseille Observatory, "that it has not been found by the civilian and military expeditions." He became convinced that investigations were being blocked by a "conspiracy" of silence among the Bedouins. The army officer was "absolutely certain that the meteorite is known to all the village natives and, more particularly, to the nomads." Like the Greenland Eskimos, "Arab blacksmiths regularly collected iron from it for working."

The "conspiracy" has proved successful, for the Great Chinguetti Iron apparently lies buried among the Sahara's shifting dunes, which can pile 1,000 feet high. During the past half-century, little has been written about the missing meteorite: most catalogues ignore its existence, listing only the 10-pound specimen returned by Captain Ripert.

Another space-oddity distinguishes Mauritania. A nation about half of Mexico's size, she counts 10 meteorite craters, while the remainder of Africa contains not half that number. (Despite Mexico's largess of iron meteorites, she bears no star-wounds.) The first Mauritanian crater, measuring about 800 feet wide, was found in 1950, about four miles from the suspected location of the Great Chinguetti Iron. If other asteroids exploded against the desert floor here, why was the Chinguetti meteorite permitted a soft landing?

Although modern investigators may deny the Great Chinguetti's existence, the other Hermit Kings are spaced around the world. To the unsuspecting eye, the map shows them scattered haphazardly among the land masses from Africa to China, from Greenland to Australia. With so many stones and irons herded together in Farrington Circle, Kansas Collection, and Iron Alley drop-zones, the positions of these king-size meteorites do appear reassuringly normal.

But *are* they geographically innocent? By conforming to the expected, are they perhaps hiding the unexpected, much as the multihued dots of a color-blindness test can conceal a question mark? This all-too-even distribution invites both suspicion—and experimentation. What would happen if lines were drawn on a world map through the 7 sites: lines running east and west, north and south? Would the resulting 36 rectangles be random in size and position, or would a hidden symmetry emerge?

But such a calculation must wait for now, because we will want to introduce still other coordinates. First we must ponder a new set of enigmas posed by giant meteorite craters or astroblemes (from two Greek words meaning "star-wound"); a topic that is most disturbing to anyone concerned about our globe's vulnerability to outlaw asteroids.

Part Three
STAR-WOUNDS

Chapter Ten

EXPLOSION CRATERS
Lunar Impacts ☆ *First on Earth* ☆ *Arizona Enigma*

Meteorites are "dynamite," packing more blasting potential than any known chemical explosive. Even space-babies can deliver A-bomb wallops, for their speed is their power. Take an object moving at a mere 1 foot per second, and boost its velocity to 5,280 feet—or 1 mile—per second. We might assume that its energy would be 5,280 times greater. Instead, it catapults to a force *28 million times* more powerful.

A tiny 1-ton asteroid moving at 5 miles per second (or 18,000 miles an hour) vaporizes on impact with a force equal to 2 tons of TNT. Increase that mass to a million tons and the velocity to 50,000, even 90,000 M.P.H., and its devastation can be likened to that of atomic devices.

For evidence, consider our airless and ill-starred neighbor, the Moon! A third of a million craters measuring a mile or more wide have been counted on her front side alone, and some are of enormous dimensions. Clavius Crater alone measures 144 miles across. Mare Imbrium measures 650 miles across, and originally may have been 50 miles deep. That billion-H-bomb-asteroid nearly cracked the Moon in half, while upheaving mountains tall as our Rockies and our Alps.

Today, while Earth is but tickled by a few falling meteorites, the Moon is apparently subject to continuing bombardment. Scientists are well aware of "surface transient phenomena" within several Moon craters where unexplained lights have been recorded flickering and flashing. They definitely are not volcanic. Dr. Elbert A. King, author of *Space Geology*, comments about "lunar bright spots and color phenomena... confirmed by several observers." Since medieval times, more than 800 transient phenomena—unexplained lights—have been observed on the lunar face. Some scientists interpret these curious flashes and glows as escaping interior gases becoming luminous under the intense solar radiation. A fair share of these red-yellow flares, however, must result from impacting asteroids—as evidenced by an 800-year-old account.

On June 18, 1178 A.D., according to a Canterbury, England, chronicler, "five or more men who were sitting there facing the Moon . . . [and] are prepared to stake their honor on an oath that they have made no addition or falsification . . . Now there was a bright new Moon, and as usual in that phase its horns were tilted toward the east; and suddenly the upper horn split in two. From the midpoint of this division a flaming torch sprang up, spewing out, over a considerable distance, fire, hot coals, and sparks. Meanwhile the body of the Moon which was below, writhed, as it were, in anxiety, . . . throbbed like a wounded snake. . . ."

In terms of lunar latitude and longitude, the medieval description places the event near 45°N and 90°E. Jack B. Hartung of the State University of New York points out that a 6-mile crater, Giordana Bruno, does exist in the vicinity. That and other evidence leads him to conclude that a meteorite impact was observed in 1178.

With perfection of the telescope, additional transient phenomena were noted. Spectacular sightings followed during the last century. They continue today. In April 1948, English astronomer Francis H. Thornton saw in the crater of Plato "a minute but brilliant flash of light," and compared it to an anti-aircraft shell about ten miles distant—not a bad description for an impacting asteroid.

This theory is reinforced by NASA instruments left behind on the lunar surface by Apollo astronauts, which have recorded telltale aberrations in the Moon's movements. On October 1, 1977, NASA pushed the buttons shutting down by remote control its instrument network on the lunar surface that had been measuring moonquakes and meteorite impacts. During its 7-year operation, ALSEP (Apollo Lunar Science Experiment Packages) beamed startling information back to the Johnson Space Center near Houston, Texas. During that period, while our globe felt only the mosquitolike prick of the Kirin City fireball, the Moon was pummeled by 1,700 asteroids hefty enough to affect NASA instruments. The largest hit in May 1973, with scientists estimating that a 1-ton projectile carved out a 300-foot crater. (Timetables for these 1,700 are interesting. The most severe lunar batterings occurred during April, May, and June, duplicating the seasonal patterns of harmless meteorite falls on Earth.)

The NASA census agitates the question: how is our planet being spared asteroidal dangers? Taking 240 lunar impacts during an average year, our little neighbor has been struck 240,000 times in the last thousand years—Earth, hardly at all.

How fortunate, we may muse, that our globe is swaddled in a security blanket of air! Devoid of any gaseous shield, the ravaged lunar surface must take all space-projectiles at full speed. But our atmosphere constantly defuses potential space-bombs, vaporizing small particles on contact, pressuring heftier missiles to detonate themselves harmless miles above the ground. More importantly, our dense air tames cosmic velocities, so that even the most massive meteorite confirmed by science— the 60-ton Hoba West iron of the Kalahari Desert—landed so softly it barely bruised the limestone beneath. With all its sound and fury, the Sikhote-Alin fireball of February 1947 lost some 200 tons by ablation, and with its speed braked by the atmosphere, was powerless to explode. Before reaching us, any incoming asteroid must penetrate the equivalent of four feet of solid rock. How reassuring!

And how foolish to believe so! "Our planet is protected by an atmosphere." The argument is immediately countered by our globe's size—a target four times broader than the Moon—and by the knowledge that air is powerless against giant asteroids.

As Brian Mason reminds us in his authoritative *Meteorites,* ". . . low terminal velocities apply only to meteorites in the size range commonly found on earth, that is, not more than a few tons in weight. When the masses become appreciably larger, conditions change drastically."

The bigger the meteorite, the less resistance our vaporous atmosphere offers. For example: an object 1 foot in diameter—about basketball size—will be virtually halted by air pressure, striking the ground at a few hundred feet per second. Make the ball 10 feet wide: its weight multiplies a thousandfold but the airbraking surface, only a hundred. Raise the speed and the weight, and you lower the barrier—with catastrophic results. A barn-size asteroid streaking into our air space at 90,000 M.P.H. will be retarded only slightly, detonating at 65,000 M.P.H. Writes noted scientist Dr. Robert S. Dietz in "Astroblemes," August 1961 *Scientific American*:

> Because of its size, a big meteorite loses little of its enormous energy to deceleration by the atmosphere. The shock that it generates upon impact must therefore transcend that of any other earthly explosion, natural or man-made. It can be calculated that such impacts produce

pressures of millions of atmospheres [one atmosphere is about 15 pounds per square inch]. Volcanic explosions, in contrast, involve pressures of hundreds of atmospheres.

We are just as vulnerable as the hammered and dented, cupped and cratered Moon. Some scientists believe that Earth has been similarly bombarded over the eons but that most of our grand cavities have been erased by the erosion forces of rain, rivers, and flood, silting, glaciers, and crustal upheavals. But not all. Our planet still bears scars by the dozens as proof.

FIRST ON EARTH

Located between Flagstaff and Winslow, 6 miles south of Route 66, Interstate 40, and variously known as the Arizona Meteor Crater, Canyon Diablo Crater, and Barringer Meteorite Crater, this world-famous cavity embraces an unsolved mystery reaching around the Earth—and to the Moon.

From the highway you first see a low butte, purple-gray in the distance, breaking the horizon of the desert flatland. Driving to the ranch-style brick museum building, you feel the blacktop road leading slightly uphill. Inside the air-conditioned museum and observation house, arranged with glass cases displaying meteorite specimens and shelves bedecked with souvenirs, four panoramic windows gaze down upon what has been called North America's greatest scientific marvel. There is the sensation of standing on a hilltop. Beyond the picture-window, the boulder-littered crater wall drops away steeply, flattening below into a plain. Roofs of tiny sheds glint in the sun. Tiny from this height, they once sheltered drilling machines. In the pinkish haze of the Arizona afternoon, the curving wall across the chasm appears streaked. Those are gullies, washed out by 20 centuries of erosion.

In size, the great amphitheater measures 4,200 feet wide and 570 feet deep. If the Washington Monument were placed in the center, its top would not poke above the rim. The huge basin would accommodate two dozen Rose Bowls and seat 2½ million spectators, or more than Arizona's entire population. You are told that a footpath encircles the crater: a hot, strenuous three-mile tramp. Any stalwarts hiking the far side are not distinguishable to you without binoculars. The rim alone—that sloping hill you drove up—is about 180 feet high and contains some 300 million uplifted tons of crushed and powdered rock.

As you realize that this open wound in the desert floor is not the product of eons-long "natural" erosion like the Grand Canyon, you begin to sense that forces undreamed of exist out there in the universe. This sunken emptiness was created in a devastating split-second by a native of deep space. Next to you, a couple stroll laughingly up to the window: he in blue polka-dot leisure suit; she in green sunglasses and pink shirt. They stop, silenced. "Heaven forbid," she whispers. "Good God," he gasps.

You choose not to ponder the fate of any luckless town or city on target for future cosmic bombs.

One question visitors often ask has yet to be satisfactorily answered: What happened to the meteorite? Was it destroyed when it hit, or is part still buried here? Replying most positively, David Moreau Barringer devoted 30 years of his life in the search for cosmic treasure. Though his project proved a financial failure, Barringer made geological history.

Curiously, giant meteorite craters were recognized first on the Moon, *then* on our globe. As early as 1826, the German astronomer von Gruithuisen proposed that certain lunar cavities resulted from impacting meteorites. Undermining his credibility, however, was the Munich scientist's earlier belief that several moonscape features were artificial, enormous structures erected by a race of "Selenites." Charles Fort's *New Lands* reports, "In the year 1821, Gruithuisen announced that he had discovered a city of the moon. He described its main thoroughfare and branching streets. In 1826, he announced that there had been considerable building, and that he had seen new streets." Toward the close of the nineteenth century, as asteroid sightings mounted, numerous scientists including the highly esteemed Dr. Grove Karl Gilbert, twice president of the Geological Society of America, were agreeing that the Moon had been bombarded by space-projectiles. But therein lay a quandary: if meteorites had blasted craters on our little neighbor, why not upon our planet too? At the time, no impact structures had been identified here.

In 1891, Dr. Gilbert visited the almost inaccessible desert crater in northern Arizona and announced that it was volcanic—the result of a steam explosion. He dismissed as coincidence the tons of meteorites that curio seekers, cowboys, and mineral dealers had been lugging away over the years from the same area. Because of his credentials and clout, Dr. Gilbert was not challenged to produce evidence of his volcano theory, so it held fast for 35 years. (Ironically, Gilbert was appointed president of the American Association for the Advancement of Science in 1900.)

Fortunately for scientific advancement, one man ignored the official edict, looking instead upon the unexplained hole in the ground with rock-hard logic. Five feet, nine inches tall, a muscular 200 pounds, a partaker of good cigars and good whiskey, David Moreau Barringer had been trained as a lawyer to gather evidence. He then became an expert geologist and ended up a no-nonsense, highly successful mining engineer. Seeking his fortune in the early 1890s, Barringer had explored scores of mines in Spain and South America, finally making a strike near Pearce, Arizona. His rich Commonwealth Silver Mine allowed him to dig into a unique commercial venture. While working as a consulting geologist in the Southwest, Barringer—with reasoning sharp and simple as a diamond drill—concluded that the "volcanic" pit was a meteorite crater.

No one could explain the cavity, measuring nearly a mile wide, as the result of anything but some kind of explosion. More meteorites had been found on a square mile of desert here than anywhere on the globe. Since the Earth contained some 57 million square miles of land area, chances favoring a meteorite blast were 57 million to 1. Furthermore, his investigation showed that the meteorite fragments had arrived simultaneously with the explosion. It was ridiculous to assume that any natural disturbance such as a volcano would occur at the same point on Earth and at the same time an asteroid detonated. The hole had to be a giant meteorite crater.

Barringer was also convinced that the impact mass, whether a single body or cluster, still lay buried under the crater floor. He calculated its weight at 15 million tons. Here was a lode of iron, nickel, cobalt, platinum, rich enough to make his silver mine look like a penny-ante operation. With the United States barren of nickel deposits, that metal alone would justify a major mining project.

After lengthy study, Barringer assumed that the meteorite had plunged straight down from the sky. He began drilling near the crater center. Core samples revealed shattered rock strata down to a thousand feet, then undisturbed sandstone. (The explosion, therefore, had not originated underground as with a volcano.) Encouraged, he kept probing for the meteorite, investing his own capital and that of enthusiastic subscribers. In 1904, the drill hit something that "rang with a clear metallic sound," yielding particles showing traces of nickel. He kept drilling. Twenty-eight holes and 7 years later, he still had only tantalizing bits of meteoritic iron, but not the main mass. Scientists openly ridiculed his venture, and his chief financial partner quit.

Realizing that something was wrong, Barringer made a simple experiment for locating the hidden asteroid. Did the round hole necessarily mean that the meteorite had dropped vertically? Taking a rifle used in big-game hunting (a sport he shared with Teddy Roosevelt), Barringer fired into some mud. Although held at an angle to the ground, the bullet made a round hole, not an oval one as expected. Other bullets, fired at sharper angles, made circular holes.

This phenomenon can be illustrated with the pebble-in-a-pond analogy: regardless of the stone's angle or direction of fall, wavelets radiate away in a circle. Unknown to Barringer, laboratory experiments in the United States and Germany using cement powder had yielded the same result. As Dr. L. J. Spencer, who investigated the Wabar craters in Arabia, had noted in his report, "Such an explosion crater will be circular in outline, whatever the angle of approach of the projectile."

Barringer suspected that the meteorite had arrived on a slant, burying itself beneath the crater's unraised southern rim. He abandoned operations on the floor, trucking his equipment atop the wall. Now, however, his drills had to cut through nearly 600 feet of additional rock: a costly change in strategy, true, but with the price of nickel pushing $60 a ton, and iron's value increasing during these boom times, he was looking at a potential billion-dollar bonanza. Barringer's own fortune, however, was nearly exhausted. Despite the bright promise below, Barringer needed 10 years to raise the necessary capital.

In 1919, drilling resumed on the southern rim. Four years later nickel particles were being returned from a depth of 1,100 feet. Barringer's enthusiasm is revealed by his operation's log: "The expert drillman says he has drilled in all sorts of formations but has never encountered anything like this. From the sawtoothed appearance of the drill bits, he says we must be passing through streaks of solid metal." At 1,376 feet the drill hit more "meteoritic material containing nickel and platinum," convincing Barringer that he at last had struck the meteorite's major body.

His feeling of elation was tragically fleeting. The drill jammed. Water filled the shaft. Barringer and his crew battled the flood for a year. Although $600,000 had been sunk into the total enterprise—perhaps equal to a couple of million dollars in today's economy—more financing was needed. Investors, however, were far from encouraged by the geological fraternity's continuing insistence that the crater was volcanic. Dr. Nininger observes, "By this time [1923] opposition to Barringer's idea had developed into a sort of tradition. To oppose was orthodox, to endorse was unsound. Consequently the opposition spoke fearlessly, and though Barringer continued to insist that his case had been proven, even his colleagues said little in his defense."

What finally killed further attempts to locate the celestial bonanza—and perhaps Barringer himself—was not scientific scorn but mathematics. In 1929, Gilbert had been dead eleven years. With most geologists at last accepting the chasm as the first meteorite crater recognized on this planet, Barringer engaged a highly qualified specialist, Dr. F. R. Moulton, to assay the size of the buried meteorite.

Behind a handsomely boyish face, Dr. Moulton hid a keen mathematical mind, expert in astronomy, celestial mechanics, physics, and ballistics. His equations encompassed the projectile's probable speed, the resistance of the rock strata, temperatures and pressures generated by the asteroid's impact, and energies required to displace some 300 million tons of earth. This was a formula for disaster to Barringer. While he had assumed that a portion of perhaps 15 million tons remained intact, Moulton stated 3 million at the most; perhaps as little as 5,000 tons. "The energy liberated in a tenth of a second," Moulton later reported, "must have been sufficient to evaporate both meteorite and the substance it penetrated, and here a mighty explosion must have occurred during which the round crater was formed without reference to the direction of impact. The crater would be the only indication of the cosmic process that had here taken place."

Although Moulton never visited the crater, and although other mathematicians challenged his conclusions, the putative prize was so shriveled that Barringer's investors departed. The 1929 stock market crash soon followed. Only 3 months after receiving Moulton's opinion, the usually robust Barringer suddenly took ill and died, still believing that he had struck a vast cosmic lode.

A failure in one sense, Barringer was undeniably successful in that he alone forced the scientific community to reject blind bias and accept evidence that the Arizona cavity was the first authentic meteorite crater recognized on Earth. The Meteoritical Society officially changed the crater's name in 1950 "to the Barringer Meteorite Crater in tribute to the pioneering work and in memory of the late Daniel Moreau Barringer, who originally proposed and ably defended the meteoritic hypothesis of the origin of the Crater and who was largely responsible for the acceptance of that hypothesis." Still, the "official" name is seldom used in scientific treatises and the man's name only infrequently mentioned in connection with the world-famous meteorite crater. (Rarely do you read accolades as voiced by the noted Russian scientist, E. L. Krinov: "As a tribute to one who had given so much time, energy and thought to its study . . . It seems only fitting that this most striking phenomenon should bear his name and be a memorial to this remarkable man.")

But was Barringer correct in believing that a huge block of metal remains interred in the Arizona desert? Or was Moulton right in calculating that the asteroid vaporized?

Scientific conflict endures; the consensus favors Moulton, not Barringer. But no unequivocal answer is possible because of one unknown factor: the meteorite's velocity. Without knowing the asteroid's speed, its size and blasting potential cannot be accurately computed.

According to Dr. E. M. Shoemaker, intimately familiar with nuclear explosives, the asteroid may have been as small as 80 feet wide, weighed a mere 65,000 tons, and struck at only 35,000 M.P.H. The scenario is not difficult to envision.

A house-size asteroid—80 feet thick, attended by a swarm of lesser meteorites, impacts at 35,000 miles per hour. Unable to escape the projectile, the upfront air compresses to the ignition point: a cylinder of the sky is afire, stabbing the target site below. Just before the alien asteroid touches the resistant Earth, the heat flares outward across the Arizona desert, searing every living thing for perhaps 100 miles. The irresistible force greets the immovable object; the interface between missile and target converts solid matter into radiation. Knowing only the airless raceway of deep space, the asteroid bores through solid Arizona bedrock for hundreds of feet before its forward portion slows abruptly. The back end continues at 35,000 M.P.H., and it turns itself inside out. As its face vaporizes, the bulk of the mass liquifies.

The meteorite's core tunnels 250 feet deep into Arizona bedrock. Sixty thousand tons of metal are converted into a gas registering a million degrees F. and exerting a pressure of 150 tons per square inch on the confining sandstone and limestone strata. A cloud of gaseous metal shoots 5 miles high; heat and shock waves stab deep underground. Subterranean waters become superheated steam and balloon in all directions: down, to the sides, and above. The rock ceiling weighing 300 million tons lifts back like an escape hatch. In an instant the Earth's crust vomits one-third the quantity of all material excavated for the Panama Canal. Vented, the metallic vapors shoot miles into the atmosphere. The mushroom cloud congeals into tiny pellets.

The dust clears. The desert floor ceases its trembling. For evidence of this cosmic encounter, there remains at ground zero a yawning chasm nearly a mile wide and a thousand feet deep that once contained 300,000,000 tons of solid rock.

Although erosion over the centuries has slightly modified its dimensions, the northern Arizona crater remains an awesome testament to the explosive power of a "baby" asteroid. Dr. Shoemaker compares it to a 1.7 megaton explosion: 1.7 million tons of dynamite. Yet even David Moreau Barringer, who invested much of his personal fortune fruitlessly drilling here for nickel and iron, remained unaware of the chasm's curious geometry.

Most scientific comments either ignore the cavity's unnatural outline or vaguely refer to it as "almost round," or "nearly circular." Moulton thought it was round, Barringer thought it was round, and a circle it should be, according to the ballistics of a point-source explosion. The crater's true shape, however, went unnoticed until some 20 years after Barringer's death, when an aerial photograph happened to be snapped directly above the cavity. The picture revealed that the crater is a *square*— with rounded corners.

It is the awesome power of an impacting asteroid that makes the Barringer Meteorite Crater's mysterious contours so unsettling. (See Fig. 7.)

A square meteorite crater in northern Arizona? As explanation, geologists lamely refer to lines of weakness in local rocks that channeled the explosion—which is akin to channelling a hurricane with tissue paper. Or how do you blow up a square balloon?

If the Barringer square were the only noncircular meteorite crater on Earth, almost any geological rationale could be tolerated. But other cosmic cavities reveal even stranger geometries. Australia's Wolf Creek

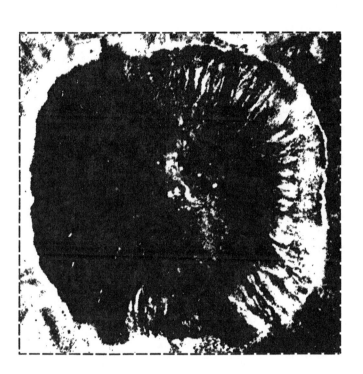

Figure 7. **Barringer Crater's Odd Geometry**   Blunted corners mark this schematic outline drawn from a U.S. Air Force serial photograph of northern Arizona Meteorite Crater. Krinov's *Giant Meteorites* terms it "an almost perfect square."

crater is prism-shaped—"markedly polygonal," observes McCall; as is the Pretoria crater in South Africa. Impact pit number 10 in the Henbury field is "markedly rectangular," states Milton. In his authoritative *The Measure of the Moon*, Baldwin says, "Often the terrestrial meteoritic craters are polygonal rather than truly circular." Quebec's 40-mile-wide Manicouagan crater is "distinctly hexagonal." You could drop a million pebbles into a million ponds and never see ripples forming such a six-sided figure upon the water. How in the world can an exploding asteroid do it, not only here but on the Moon? "Many [lunar] craters are quite clearly polygonal in shape," reports the *Larousse Encyclopedia of Astronomy*.

One "unthinkable" explanation for the Moon's many anomalies recalls von Gruithuisen's Selenite City, a notion no more popular to orthodox minds today than in 1821. *Somebody Else Is on the Moon* is an exuberant exposé of strange structures noticed by author George H. Leonard, who microscopically studied hundreds of photographs returned by Lunar orbiters. Moon cavities receive his special attention. "Some craters are octagons; some are hexagons; a few are squares." Leonard asks, "Is there a natural way in which an octagon can be formed on the Moon—an octagon thirty-one miles in diameter?" In all seriousness, he identifies gargantuan excavating equipment in the photographs. "There are rigs on the Moon several miles long, capable of demolishing the rim of a seventy-five-mile-wide crater in the same time it takes us to level ten acres."

In reading the likes of *Somebody Else Is on the Moon*, the most blissful stance to adapt is one of kind jocularity. Except that Charles Fort's news clips from the scientific press of the last century glint with observations of lunar "beacon lights." Are they the headlights of giant bulldozers carving out hexagons and squares upon the Moon's virtually unexplored surface?

The only possible eyewitnesses to the Barringer asteroid were early Indians, and their legends make no mention of "machines" excavating the crater. According to Dr. Nininger, who lived in the area, Indian tradition speaks of a Great Spirit who made a fiery descent from heaven to disappear beneath the desert. *National Geographic* magazine (June 1928) cites a Hopi Indian legend detailing how a trinity of gods arrived there—one using the crater as his temple-abode; the other two landing some distance north. Both legends obviously refer to blazing asteroids; a fairly well documented date places the event at 22,000 years ago, again demonstrating the durability of tribal memory.

Legend seems unaware, however, of a similar phenomenon occurring at about the same time 550 miles southeast, near today's Odessa, Texas. A mini-asteroid weighing perhaps 300 tons impacted the mesquite-covered desert, blasting away 100,000 tons of sand and stone, leaving an explosion crater plus several impact pits.

In 1926, Barringer's son visited the site, reporting: "It is a meteor crater, beyond the shadow of a doubt." He described it as oblong, about 450 by 600 feet with a rim of broken rock between 16 and 18 feet above the crater floor. His father assumed that yet a second cosmic treasure lay beneath the Texas plain near Odessa, and attempted to buy the land at the going rate of $7.00 an acre. But with the discovery of oil and gas deposits nearby, the pricetag went out of sight. Other scientists visiting the craters confirmed their extraterrestrial nature, and Odessa became the second astrobleme recognized on Earth.

In 1942, perhaps as omen of things to come, a local newspaper, the *Odessa American*, reported that the basin offered "enough territory to hold 10 large oil storage tanks." The craters served as an industrial dump instead. Civic pride temporarily forestalled their fate when the chamber of commerce envisioned a "Shooting Star Park," featuring the world's first subterranean meteorite display. An elevator would whisk paying tourists 160 feet down, where a catwalk would allow them to view the freshly polished metal mass, believed to be 18 feet tall. But the growing realization that asteroids disintegrate on impact, plus the urgencies of World War II, aborted hopes for the underground attraction.

Time and Texans have not been kind to the Odessa craters. With uncharacteristic emotion, Dr. Ralph B. Baldwin lamented in 1963, "It is a crying shame that the county of Ector or the state of Texas has not done something to preserve this ancient group of craters. They are now being overrun by an oilfield and are being virtually destroyed by dumping and digging." Another concerned scientist suggested changing the crater name to "Odessa-Cration."

In 1979, conditions there had not improved. Although a sign on Interstate 20 brightly alerts motorists "Meteor Crater 7 Miles," only a rusted-out sign marks the 2-mile gravel access road. It leads past mesquite trees and wind-tossed tumbleweeds on one side, and on the other past an oil field with pump-jacks nodding down and up—thirsting iron birds at drink—to the Odessa site. The second meteorite crater ever identified on earth stands forlorn and neglected, looking like an abandoned land-fill. Barely 5 feet deep now, the weedy basin is torn by motorcycle tracks and littered with beer cans and a washtub. Little rodents scurry behind the jagged rocks marking the nearly leveled rim. A garage-size structure—perhaps Shooting Star Park's museum?—awaits ultimate decay. Winds hustling through broken windows stress the feeling of empty sadness. One of our globe's rarest phenomena belongs not to science any longer, but only to the nodding pump-jacks, to the scurrying rodents, and to the west Texas wind. In April 1979, according to the *Odessa American,* the city's Convention and Visitors Bureau suggested the site be returned to its natural state because "It is no longer definable as a meteor crater. Maybe we could pave some road out to it."

From earlier examination, many authorities suspect a physical linkage between Odessa's mini-astrobleme and the Arizona star-wound. Their ages are approximately the same. In both instances, the incoming asteroid evidently arrived from the north. Odessa fragments are "essentially identical" to Barringer's. One conclusion has a parent body separating in the atmosphere—the major portion impacting in Arizona, the other in west Texas.

After the Odessa craters, other discoveries soon followed: Campo del Cielo, Argentina; Henbury, Boxhole, Wolf Creek in Australia; Wabar, Arabia. Geologists seldom comment on why so many craters measuring less than a mile are located in the world's deserts. But then, these desert craters present mysteries all their own.

Chapter Eleven

A COSMIC GEOMETRY

Desert Dilemmas ☆ *The Super-Craters* ☆
Rectangles in Time

The world-count of explosion craters measuring less than a mile wide runs between 8 and 10. Identification is often difficult due to great age, erosion, and the loss of telltale meteoritic material following the explosion. Excluded from this investigation are impact pits like Campo del Cielo, Argentina; Haviland, Kansas; Sikhote-Alin, Siberia, and Monturaqui, Chile—because their asteroids evidently *shattered* rather than detonating on contact. In *Space Geology*, Dr. Elbert A. King lists 9 minicraters (with exceptions noted) measuring less than a mile across. Reminiscent of the Hermit Kings' regal isolation, all but one crater is found in the world's deserts: since publication of King's 1976 list, satellite photographs evidently confirm a tenth crater in lower Mongolia, inscrutably named Tabun Khara Obo.

The Southern Arabian Desert between the Persian Gulf and the Red Sea holds the remote Wabar or Al Hadīdah duet of craters. (Once again, legend preceded discovery, and a baking unwelcome awaited the visitor to these minor astroblemes.) The Texas-broad thirsty wilderness of the Empty Quarter, containing one of the largest sand areas in the world, may go rainless for 20 years.

Today's aerial navigation charts show a stippled blankness, indicating desert. Map-notes read laconically, "sand dunes," "undulating plain," "sand and rocks," "no vegetation." Bedouins warned the intrepid English explorer Harry St. John Philby, "Where there is no water, that's the Empty Quarter; no man goes thither." Another Englishman, T. E. Lawrence ["of Arabia"], wrote in 1929, "Only an airship could cross it."

Though the permanent population remains zero, the Empty Quarter is untraversed no longer, thanks to the liquid magnet of Arabian oil and four-wheel-drive vehicles. Geologists and technicians rove the desolate sands in diesel caravans of air-conditioned, two-story mobile homes equipped with living rooms, kitchens, and—the desert's ultimate luxury—

showers. Provisions still must be air-lifted, however, since the Empty Quarter remains lifeless.

But in 1932, Philby had "worked and waited" fourteen years to track down a strange Bedouin tradition: somewhere out in the desert lay a ruined city called Wabar, "destroyed by fire from heaven" because its wicked Persian ruler refused to honor the Prophet Hud. Legend also confided that near the site lay a block of iron "big as a camel's hump" that tribesmen called Al Hadīdah. Knowing that *hadid* meant "iron" in Arabic, Philby presumed that there were iron artifacts at the lost city. Adapting the Arabic language, Moslem religion, and Bedouin garb, Philby yearned to be the first European conqueror of the Empty Quarter by a successful crossing. That honor was denied him however: one year before his expedition began, his compatriot Bertram Russel traversed the Ar Rab' al Khālī. Philby further failed in his quest for Wabar, yet his perilous trek into the Empty Quarter proved astonishingly successful for unexpected reasons.

From the Persian Gulf coast in January 1932, Philby led an expedition of 18 Arab tribesmen and 32 sand-bred camels into "an ocean of sand," he later wrote in the British *Geographical Journal*, "of which the ridges were the waves." Unbearable heat and lack of water forced half the party to turn back. Cheered and chided by Philby, the remainder pushed on. "Even the raven knew it was no use looking for food here." Two camels were slaughtered for their meat, a third went lame and was abandoned. During one stretch, the animals went 2 days without food, 10 without water. Wisely the Bedouins prevented the suffering beasts from drinking, instead pouring the precious water slowly down the camels' nostrils "to cool their brains."

Despite all hardships, Philby carried on in the finest British tradition. "I did 55 days without a cold drink of any kind. My substitute being tea and freshly drawn camel milk, warm and frothing." The explorer spent 3 months covering 1,800 miles of the Southern Desert, voicing only a small complaint: running short of tea, he had to use the same soggy leaves "over again for three or more brews."

Philby's guides did not share his grand determination to find Wabar and Al Hadīdah, but he enticed them with rewards: $30.00 for the great hunk of iron; $5.00 for every building discovered. At last the thirsting band stood on a low dune. With surprise and disappointment Philby "looked down, not at the ruins of a city, but into the open mouth of what

I took to be a volcano with twin craters side by side, surrounded by low walls of what looked like outpoured slag and lava." Perhaps rationalizing that an "extinct volcano" could be mistaken for a desert citadel, he described a structure "whose black walls stood up gauntly above the encroaching sand like battlements and bastions of some great castle." His disappointment deepened when, instead of an iron mass big as a camel's hump, he uncovered "a silly little fragment . . . about the size of a rabbit" a meteorite specimen weighing 25 pounds. Their reward money uncollected, the Arabs eagerly stuffed their saddlebags with "little jet-black shining pellets" they believed to be pearls—but were, in reality, only worthless globules of glass.

Philby was unaware that he had discovered a group of meteorite craters. At the time, only 6 others were known in the entire world. Scientists today count 2, possibly 3, astroblemes at Al Hadidah. The largest chasm measures about 300 feet wide. Its sides slope down some 40 feet. Before sand began filling the basin, it may have been twice as deep. Within historic times, a blazing asteroid evidently separated before impact; its fragments exploding in the deep, loose sand within a quarter mile of each other. The impact blasted molten sand into the atmosphere, where it congealed, drizzling down as glassy pellets. Dr. L. J. Spencer, formerly Keeper of the Minerals at the British Museum, explains, "The kinetic energy of a large mass of iron [nickel-iron] travelling at a high velocity, was suddenly transformed into heat, vaporizing a large part of the meteorite and some of the earth's crust, so producing a violent gaseous explosion, which formed the crater and back-fired the remnants of the meteorite. Such an explosion crater will be circular in outline whatever be the angle of approach of the projectile. The materials collected at the Wabar craters afford the clearest evidence that very high temperatures prevailed. The desert sand was not only melted, yielding a silica-glass, but this boiled . . . and was vaporized. The meteoritic iron was also in large part vaporized, afterwards condensing as a fine drizzle. . . ."

The fact that the meteorite did not vaporize entirely again demonstrates the Bedouin legend's accuracy, but confirmation for a mass of metal big as a camel's hump had to wait more than 40 years. Writing in the January 1966 *National Geographic* magazine, author-photographer Thomas J. Abercrombie recounts his discovery "Beyond the Sands of Mecca." Traveling in a Land-Rover, he and his party drove "more than

400 miles with hardly a landmark." At Al Hadīdah they inspected the craters, and then, "Rumor had become reality; the biggest iron meteorite ever found in Arabia lay at our feet . . . Shaped roughly like a saucer, it measured about four feet in diameter and two feet thick at center. A little quick geometry put its weight at almost two and a half tons. . . ."

Although any meteorite traveling fast enough will detonate upon striking Earth, the cosmos shows a strange favoritism in the type that create these lesser craters. The layman's three categories of meteorites—irons, stones, stony-irons—are far too broad for scientists. They classify meteorites in technical detail, using dozens of terms like "nickel-rich ataxite" for a type of iron, and "hypersthene achondrite" for a kind of stone. Here's the mystery: no particle or fragment from the detonation of a stone or stony-iron has yet been recovered. Actual projectiles seem limited to irons. "It is remarkable," states Dr. V. F. Buchwald, of Lyngby University, Denmark, "that the eight craters were all caused by iron meteorites, and that only two types of iron were involved." Even though some craters may date back 100,000 years, Dr. Buchwald claims that the projectiles "consisted of two types known as medium octahedrites . . . and coarse octahedrites." The reason for this long-term consistency is unclear.

And iron asteroids producing mini-craters do show a curious affinity for deserts—as in Australia, where 3 explosion cavities under a mile wide have been identified.

In the northwestern part of the Great Sandy Desert, whose low dunes and loose sands are tufted with spinifex or porcupine grass, lies Wolf Creek crater. Although the astrobleme was sighted by airplane in 1937, no scientist penetrated the virtually unexplored area until 1948. They found a cavity 2,800 feet wide and 200 deep, rimmed with broken rock strata, and 3 meteorite specimens weighing about 300 pounds each. In shape, Wolf Creek crater has been described as "highly asymmetrical," an understatement worth subsequent comment.

At the central part of the Desert Continent, 180 miles apart, lie the Boxhole Station crater and the Double Punchbowl of Henbury. So similar are the chasms in size and so alike their meteorite fragments that some scientists believe that the astroblemes resulted from a disintegrating asteroid. Other authorities claim that *all* Australian craters (including the Liverpool exhibit to the north in the very undesertlike Arnhem Land) resulted from a single massive projectile. Judging from alignments of the Henbury circles, it was a most unusual missile indeed.

The Henbury field consists of 12 craters; all but 4 are evidently impact pits. Around them stretches the flat and barren desert of the "Never-Never," baked by a relentless sun and knowing barely 8 inches of rain a year, much of it pouring down in torrential summer storms. The field is situated in the dry water course of the Finke River. The largest craters act as catchbasins, sprouting grasses, shrubs, and acacia trees. In this arid land where even the hardy needlebush has difficulty surviving and any greenery is "extraordinarily scarce," the leafy tops of 45-foot trees appear like an oasis in the distance.

Closer inspection reveals a series of basins, heavily eroded and looking like abandoned gravel pits. The largest crater is a 600-foot oval some 50 feet deep that Australians call the Double Punchbowl. Geologists see it as 2 craters in 1, excavated by twin projectiles landing side by side. Adjacent lies the smaller Water Crater, a breach in whose wall admits ground water during the raining season, providing a reservoir for the occasional kangaroo or dingo. Browsing cattle have worn trails down its 20-foot slopes.

The Henbury craters' exact age is unknown. According to botanists, some acacia trees have been standing for 300 years. Scientists give the structures a tentative age of 4,700 years. The ancestors of today's Aborigines may have witnessed the asteroid's explosive arrival, for natives shun the area, calling it *chindu chinna waru chingi yabu*, "sun-trail-fire devil stone," and refusing to camp nearby.

An exhaustive study was made of the Henbury craters by Dr. Daniel J. Milton of the U.S. Geological Survey in collaboration with the Lamont Geological Observatory of Columbia University and the Australian Bureau of Mineral Resources. The 1963 project was sponsored by the National Aeronautics and Space Administration. Dr. Milton's 17-page report (Professional Paper 599-C) includes a detailed map of the Henbury field, plus the following schematic of the four largest craters. The scientists make no comment on any possible mathematical relationships among the craters' contours.

Their circles are typical of most point-source explosions created by high-velocity meteorites impacting the terrain. This family grouping, however, appears suspiciously snug, and invites further investigation.

An unexpected rectangle can be drawn touching the outside walls of all four craters. (Arrows mark the intersection between floors of the 2 craters forming Double Punchbowl.) These points within the rectangle provide intervals for a hidden grid, formed by lines dividing the rectangle into thirds vertically and horizontally. Intervals for the grid are provided by distance of Point 1 from top and side of original rectangle. Point 2 is located at another intersection—where the largest crater floor (solid circle) crosses its partner's outside rim (broken circle). At Point 3, the grid-line runs tangent to the crater floor. (See Fig. 8.)

Such aesthetics and precision do seem inappropriate for the wild, ferocious, uncaged power of space-bombs, blasting sand and rock by the thousands of tons. Why should craters interlock not just physically, but mathematically? Is a rambunctious Mother Nature that neat an architect, or is Lady Luck reaching one more time into her seemingly inexhaustible bag of coincidences?

Yet these projectiles and their tombs cannot be classified as major encounters. Listen to Dr. Robert S. Dietz's description of events creating the 30-mile-wide ring of hills near Vredefort, South Africa*:

> . . . the event that produced this structure emerges beyond doubt as the greatest terrestrial explosion of which there is any clear geological record. Apparently an asteroid a mile or so in diameter plunged into the earth from the southwest, for the structure is overturned somewhat to the northeast. The huge object drilled into the earth and released enormous shock forces, causing a gigantic upheaval. Strata nine miles thick peeled back like a flower spreading its petals to the sun, opening a crater thirty miles in diameter and ten miles deep. The shock must have reached with shattering force down through the entire 30-mile thickness of the earth's crust.
>
> . . . The Vredefort blast was . . . probably several thousand times larger than the greatest possible earthquake. In the terminology of nuclear explosions it was at least a 1.5 million-megaton event [one megaton is the equivalent to the force exerted by the explosion of a million tons of TNT]. . . .

With poetic accuracy, Dr. Dietz has coined the term *astrobleme*, "star wound," for explosion craters like Vredefort. Despite its size, Vredefort is dwarfed by other star-wounds so vast that their craters appear on a map as natural features of landscape.

*"Astroblemes," August 1961, *Scientific American*

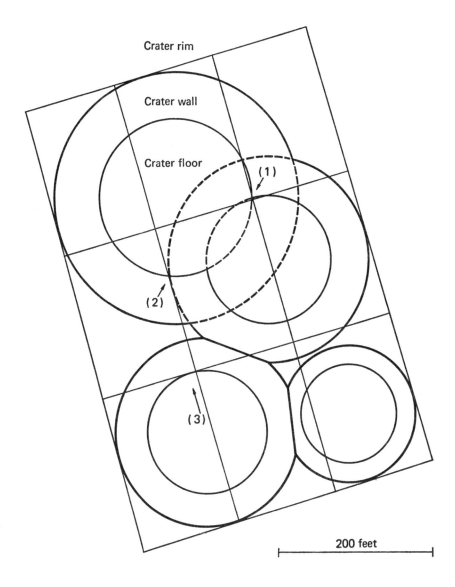

Crater rim

Crater wall

Crater floor

(1)

(2)

(3)

200 feet

Figure 8. Schematic of Henbury Craters   Broken lines indicate missing walls of two overlapping craters that form the Double Punchbowl (*after J. D. Milton*).

Some scientists point to North America's scalloped Atlantic seaboard: to the 1,000-mile-wide Gulf of Mexico, and to the cutout coasts of Florida and the Carolinas as possible astroblemes. Further north, a quartet of great bays are eyed suspiciously: the Gulf of Maine, Gulf of St. Lawrence, Ungava Bay, and the half-circle known as the Hudson Bay Arc carved from Quebec Province.

In trying to verify or discredit the cosmic origin of large astroblemes—say, those exceeding 10-mile diameters—scientists are frustrated by the meager evidence available. Most large craters are extremely ancient, millions of years old, and any fragments of the original asteroid have long since dissolved into sand, rust, and dust. Any tortured rock strata within the crater testifying to the impact often are either concealed under lake waters or by thick sediments, the outwash of surrounding walls gradually leveled by rain, rivers, even an invading sea, and, in northern climes, by mile-high glaciers. Star-wounds frequently are located in nearly inaccessible wilderness like subarctic Canada.

Sheer size can make fieldwork tedious. For example, water-filled Manicaouagan crater in central Quebec Province is rimmed with forested hills, and demands a 120-mile hike of anyone wishing to inspect its perimeter. The cost of equipping and transporting a modest expedition can be a deterrent to exploration, and investigators' results often are inconclusive.

For these reasons, no "official" list of giant meteorite craters has yet been compiled. The roster lengthens and shrinks, pulsing with geologists' efforts to confirm or deny suspected star-wounds. One catalogue shows 200 candidates, another 50, while one authority would discredit practically the whole lot, rejecting the very existence of astroblemes on Planet Earth. My primary source is Dr. Elbert King's 1976 *Space Geology*, listing a modest 39 explosion craters more than a mile wide. Despite the conservative count, the list is full of surprises.

A world map offers the first astonishment: where giant craters are *not* found. Asia—with nearly a third the Earth's land surface as a target for asteroids—is blank. India, China, and Russia are unmarred by craters measuring more than a mile wide. Recently, however, Soviet scientists reportedly have confirmed 2 star-wounds exceeding 30 miles. According to Dr. L. Firsov of the Academy of Sciences, one lies about 250 miles northeast of Moscow. Puchezh-Katunki crater, now filled by the waters of

Gorki Reservoir, was excavated some 180 million years ago by a Hermes-like asteroid that missed us in 1937. The other Soviet discovery is located within the Popigai Basin of northern Siberia. Further east, Siberia does contain the Sikhote-Alin cavities, but as in Kirin City, China, these are impact pits, not explosion craters.

Stingy numbers of giant craters mark other continents: South America, none; Africa, 3; Australia, 2. Europe, however, can claim 8. How curious that Europe's quota should surpass the combined total of Asia, South America, Africa, and Australia!

Another surprise: across the Atlantic, North America enjoys a bounty of 26—most spotted in the northeast. This continent also shows an imbalance, reversing the ratios of meteorites between United States and Canada. While the U.S. boasts record numbers of specimens, and its northern neighbor hardly a wheelbarrow full, cosmic Canada outdoes the world in giant meteorite craters. What Kansas is to large stone meteorites and what Iron Alley is to huge metal deposits, Canada is to astroblemes. Nine for the United States, 17 for Canada, and her ultimate count may reach 30, not including the super-craters of the Gulf of St. Lawrence, Ungava Bay, and the Hudson Bay Arc.

The crater picture becomes no less puzzling with mini-explosion cavities less than a mile broad. King lists 9, all but 1 located in deserts: America's Southwest, Mauritania, Australia, Arabia. A recent satellite photograph offers a new candidate in the Gobi Desert.

Astroblemes and mini-craters now will be map-inspected, yielding a last surprise that ties geography with geometry. The usually circular outlines of astroblemes have performed some geometrical gymnastics by transforming themselves into squares, hexagons, and a uniquely asethetic rectangle.

Remember the Hermit Kings? Two sets of clues flagged our suspicion. First, these behemoths seem *too evenly* distributed about the world; *too equally* apportioned among the continents, both north and south of the equator. In a map-experiment, lines are drawn east and west, north and south, through the 7 sites of the Hermit Kings. (See Fig. 9.) The lines deliver 36 rectangles of varying sizes. Only a random pattern is immediately evident. But a curious symmetry is created at the top and bottom of the diagram by 6 of the largest rectangles (see dotted lines). Is this arrangement by pattern or design?

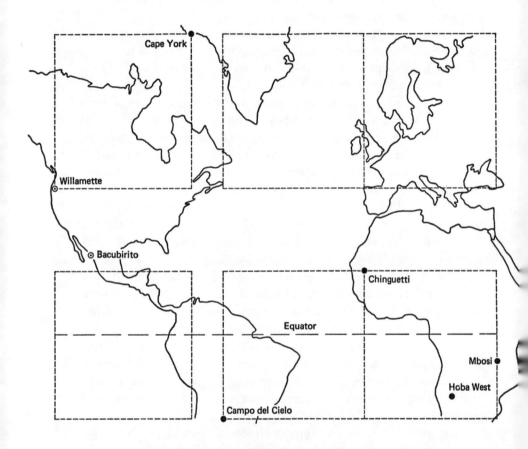

Figure 9. Rectangles formed by linking the Hermit Kings

More than semantics is involved here, because patterns frequently occur accidentally in nature. Example: wintry decorations etched upon a window by Jack Frost. Any map-patterns among the Hermit Kings are quite acceptable. But when we apply the term "design" to the 6 map-rectangles, however, we automatically assume that the Hermit Kings' landing sites resulted from an intellectual concept.

This outlandish thought seems defeated by the diagram itself. It is too selective, using only 6 of the 36 rectangles available. The precise locations of 3 Hermit Kings—Chinguetti, Campo del Cielo, Greenland—are unknown. The diagram does not depict the true world, but a crudely distorted flat projection of the globe. Therefore any symmetry among the rectangles can be dismissed as coincidence—a pattern of accident.

However, the simple geometry uniting the Henbury circles within their rectangular grid is enhanced by another mathematical nicety involving the 2 largest craters. In each instance, the rim-to-rim distance is slightly greater than 1.5 times the width of the crater floor; more exactly, the ratio is 1.6 to 1. Anyone familiar with the fine arts or architecture will immediately recognize this proportion as the Golden Section, so often applied to a painting's composition and to a building's design.

The Golden Rectangle is simple to construct. Divide a square of any size in half, into 2 equal rectangles. Then draw a diagonal through any half-square. The length of that diagonal, added to the short side of the smaller rectangle, will result in a finished Golden Rectangle. The shorter side of this attractive playing-card shape is slightly more than half, or .618, the longer dimension. Since antiquity, artists and architects have applied the Golden Proportion in their works (examples include the Great Pyramid, the Parthenon, Chartres Cathedral, paintings by Leonardo, modular industrial designs by Le Corbusier).

The aesthetic proportion is among Mother Nature's most popular and versatile design elements, and found in many living creatures. By translating a straight line into a curve, Nature achieves the Golden Spiral as fashioned into the chambered nautilus' graceful shell and the ram's artfully curving horn. Nonliving objects also sport the Golden Spiral: oceanographers admire it in the curl of a breaking wave; astronomers in the pinwheel-like galaxies of spiral nebula. One authority maintains that a meteorite digs a pit that corresponds to a segment of the Golden Spiral.

Although Nature applies the magic decimal to astronomy, physiology, and botany, it appears lacking in geology: at least I have read no mention of mountain peaks, for example, spaced at golden intervals across the land. Therefore, the appearance of .618 among the meteorite craters of Henbury, Australia, seems extraordinary. This astonishment broadens when we find it spanning the Atlantic Ocean.

Chance might be an admissible rationale if the Golden Rectangle did not reappear among Scandinavian astroblemes and among Canadian super-craters. Explosion craters huge enough to humble a hydrogen bomb seem unlikely places for more meteorite mysteries. And yet . . . !

A 2,000-mile-ocean now yawns between western Europe and eastern North America, but they are bound together by a Golden Rectangle, pegged to giant meteorite craters on either side. Lesser rectangles can be constructed within each group, pylons supporting the golden transatlantic bridge.

The 7 European astroblemes appearing in the following diagrams are taken from the *Space Geology* list. Five are lakes spotted around the Baltic Sea. Most northerly is 15-mile-long Lappajarvi (Lapp Lake), inland from the city of Vaasa. Across the Gulf of Bothnia, southern Sweden offers 3: Lake Dellen, the exceptionally beautiful Lake Siljan, and at the lower tip, little Mien, about 3 miles wide.

The fifth Baltic crater is a mini-astrobleme left over from "Desert Dilemmas." Structures like Henbury and Barringer are usually found in arid regions, and Kaalijarvi is located on an island off the Estonian coast. Barely 300 feet wide, it resembles a Christmas wreath in aerial photos: a circular pond fringed with greenery. Before World War I, the owners of the local estate built a pavilion for enjoying afternoon tea. Kaalijarvi's age is estimated at 4–5,000 years.

Southern Germany offers the Rieskessel, a 15-mile astrobleme near Stuttgart. Within its worn-down rim nestles the ancient walled town of Nordlingen. Geologists believe that the "giant kettle" was blasted by an asteroid between 15 and 20 million years ago. Last of the European star-wounds to be plotted is 9-mile Rouchechouart in southwestern France. To thoroughly scramble the puzzle pieces making up the Golden Rectangle, we arbitrarily add Opava, Czechoslovakia, where Stone Age man assembled his collection of meteorites 20,000 years ago.

In Figure 10, a series of similarly-sized rectangles mount across Europe for some 1,500 miles. Notice how Opava Stone Age site forms the right side of a Golden Rectangle. The baseline joins two extreme points:

Heralded by a thundering fireball that cracked plaster in Kansas homes, the Norton stone meteorite buried itself nine feet deep in a wheat field in February 1948. Weighing 2,360 pounds, it was the largest space rock yet seen to fall. Most curiously, Kansas counts more stone specimens than any other similarly-sized region in the world (Courtesy: Dr. C. Bertrand Schultz, former Director, University of Nebraska State Museum)

Another of the "Februarians" was the Kirin City Meteorite of March 1976. Had it exploded, rather than burrowing eighteen feet into the frozen Chinese soil, it would have undoubtedly annihilated the nearby village
(Courtesy: *Science News*)

Recovered, the sky-prize is admired by villagers. The Kirin City Meteorite weighs 3,900 pounds, the largest stone space-missile yet observed to land. Curiously, China owns only a dozen meteorites; the United States, over 700 (Courtesy: *Science News*)

The Willamette (Oregon) Meteorite, largest ever found in the United States, should have landed broad-face down, as displayed here. But for some reason it flipped in midair, landing pointed end forward, barely puncturing the soft ground beneath it (Courtesy: Oregon Historical Society, Portland, Oregon)

Today displayed in New York City, the fifteen-ton Willamette shows cavities gouged and sculpted by air friction that give it the look of a giant metallic hunk of Swiss cheese. Strangely, this huge mass of nickel-steel landed in a region already studded with giant iron meteorites
(Courtesy: American Museum of Natural History)

The search continues, as yet another treasure-hunter backpacks into the craggy wilderness east of Port Orford, in hope of relocating Oregon's eleven-ton phantom meteorite, "lost" since 1856. The scientific value of the extremely rare stony-iron specimen is incalculable
(Courtesy: *Oregon Outdoors*)

The thirty-four-ton Ahnighito iron, discovered *in situ* by explorer Robert E. Peary in 1897, was the largest of a family of three clustered in the Greenland coast near Cape York. Since then, that Arctic wilderness has yielded several other weighty metal meteorites. The cosmic popularity of northeastern Greenland has yet to be explained (Courtesy: American Museum of Natural History)

Moving the "monster" aboard ship proved a herculean engineering task. The Eskimos believed their "heavenly guest" was occupied by a demon. "There were many incidents . . . to suggest the supernatural," Peary admitted. En route to New York, his ship barely escaped being crushed by an iceberg, holed by a hidden reef, and capsized by savage Arctic gales
(Courtesy: American Museum of Natural History)

World's largest known meteorite, the sixty-ton Hoba West iron of South-West Africa, enjoyed a relatively soft landing, barely burying itself in the earth. Like so many of the "Hermit Kings," this global champion came down in a semi-desert wilderness   (Courtesy: American Museum of Natural History)

Explosion craters of giant meteors were first detected on the Moon's pockmarked face. Not until later did scientists admit that the Earth too bears star-wounds left by space-projectiles far more massive than even the hefty Hermit Kings
(Courtesy: Field Museum of Natural History, Chicago)

The three-quarter-mile-wide, 600-foot-deep Barringer Crater in Northern Arizona was blasted from the desert floor by a cosmic missile probably no larger than the museum building now perched on the left-hand rim. Within the seemingly circular outline lies concealed a strange geometry, most apparent when the crater is viewed from directly above
(Photo by Peter L. Bloomer; Courtesy: Meteor Crater Enterprises, Inc.)

Australia's Henbury Craters are a series of impact-basins, two of which overlap to form a structure called the Double Punchbowl. While not so impressive from the ground, these craters reveal a most unexpected geometry when seen from above
(Courtesy: American Meteor Society)

This cannonball, superimposed on Manhattan, suggests the bulk of asteroid Hermes (in reality, as irregularly-shaped as Manhattan itself) which missed our planet by a mere half-million miles in 1937. One scientist has estimated that Hermes' impact with Earth would have created an explosive power equal to a billion H-bombs. And since 1937, many more "outlaw asteroids" have been detected in near-misses
(Courtesy: American Museum of Natural History)

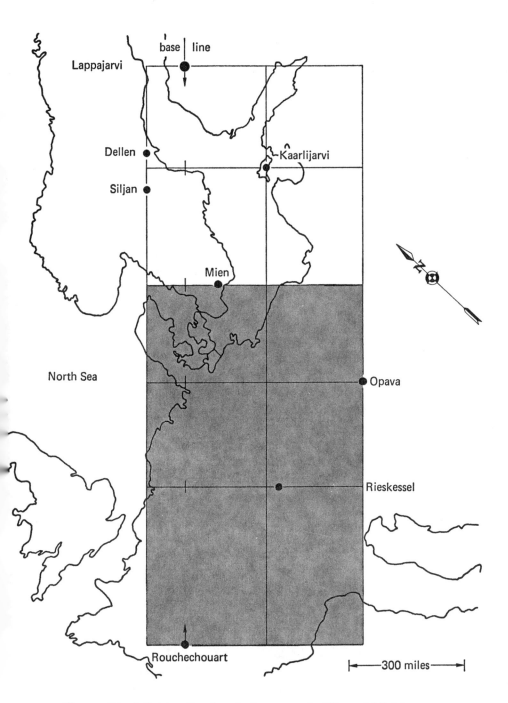

Labels on figure:
base | line
Lappajarvi
Dellen
Siljan
Kaarlijarvi
Mien
North Sea
Opava
Rieskessel
Rouchechouart
|←————300 miles————→|

Figure 10. A Golden Rectangle forms part of the grid linking
meteorites across Scandinavia and Western Europe

Lappajarvi and Rouchechouart. Twin rectangles join Siljan-Kaalijarvi-Mien. The baseline connects Mien and Lappajarvi. Interestingly, a different map projection does not deliver these geometries. In testing for interrelationships among meteorite sites, there is no "perfect" map; all projections must distort the globe in some manner. But: unlike alignments still yield Golden Rectangles.

The 4 semicircular silhouettes of East Canada have long perplexed scientists. No known geologic force on Earth—not volcanoes, not sea erosion, not glaciers—can account for their symmetry and their grand dimensions. Most majestic is the Hudson Bay Arc, a 280-mile crescent scooped from Quebec Province's western shore. To the north lies Ungava Bay, whose name in Eskimo means "far away." East lies the Gulf of St. Lawrence, the half-moon of its lower reaches formed by the curving seaboards of New Brunswick and Nova Scotia. South lie the Gulf of Maine and Massachusetts Bay.

With discovery of numerous astroblemes in Canada, and especially in Quebec Province, some scientists ventured that the 4 bays might be the ribs of unbelievably ancient star-wounds. One authority calculated that the Hudson Bay Arc's 60,000 square-mile basin would require a 21-mile asteroid traveling at 22,000 miles an hour to excavate. If authentic impact structures, the 4 must be classified as super-craters, dwarfing even the "giant" meteorite crater of 40-mile Manicouagan. Today, however, their cosmic origin is doubted. In 1972, Dr. Robert S. Dietz and J. Paul Barringer explored the Hudson Bay Arc "by Indian and Eskimo canoes and fishing boat to investigate its possible astrobleme origin." After their examination, the two concluded, "an impact origin for the Hudson Bay Arc, although unsettled, seems unlikely."

But the crescent is no cartographic illusion. Even on a wall-map of Quebec, the Hudson Bay Arc is a half-circle. Except for a fringe of offshore islands, it appears stamped from the landscape by some huge cookie-cutter. And the Hudson Bay Arc and its 3 sister bays are related mathematically by a parent circle and by a system of Golden Rectangles.

The centers of circles inscribed within the 4 gulfs supply points in diagramming the Golden Rectangles. In Figure 11, the arrow indicates the intersection of the 2 lines within Manicouagan crater. In Figure 12, the Quebec Circle, centered at Manicouagan, passes through the midpoints of Hudson Bay Arc, Ungava Bay, and Gulf of Maine. Circle's

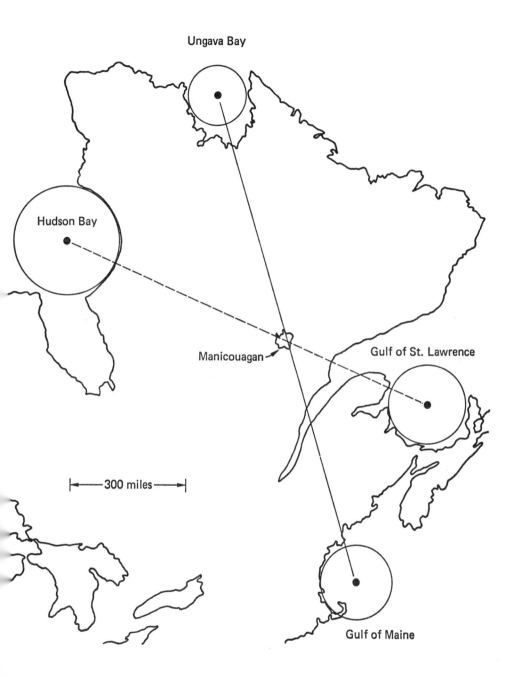

Figure 11. Four great bays dwarf the "giant" meteorite crater of Manicouagan in Quebec Province. Circle centers are marked by dots to be used in developing Golden Rectangles.

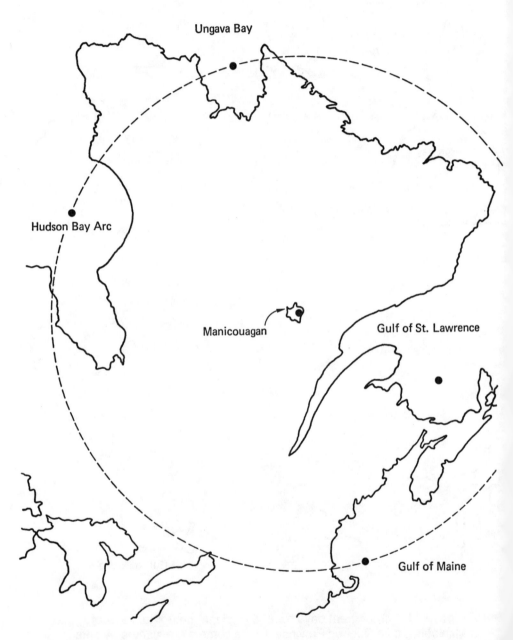

Figure 12. A circle with the Manicouagan intersection as its center precisely links the central points of the three outer bays.

Figure 13. All five points are also related by a Golden Rectangle.

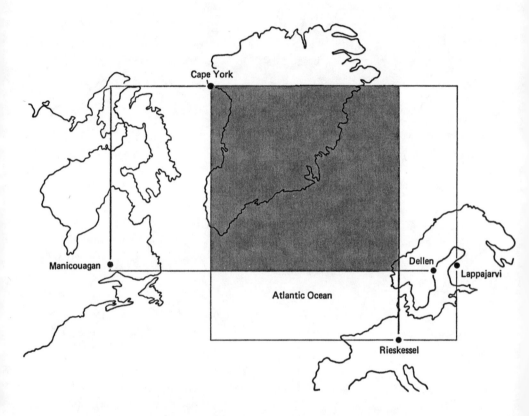

Figure 14. Trans-Atlantic base line joins Canada's Manicouagan crater and Finland's Lappajarvi. Distance: some 3,300 miles. Notice how this projection bloats Greenland, actually about Scandinavia's size. Cape York meteorites, discovered by Peary, participate in the most northerly side of a Golden Rectangle.

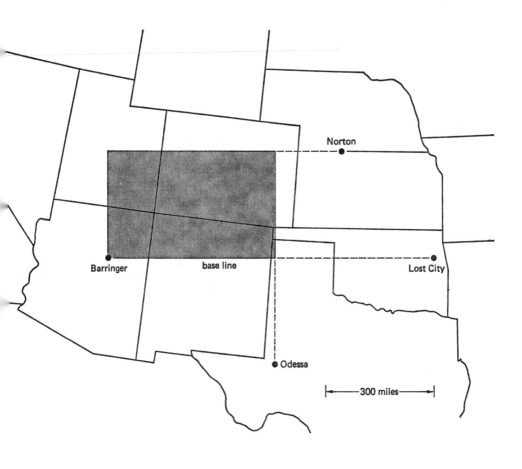

Figure 15. Yet another Golden Rectangle links the Barringer and Odessa craters to the Norton (Kansas) and Lost City (Oklahoma) sites.

center is a few miles from the intersection shown in previous figure, but within the 1 percent allowable error.

In Figure 13, a Golden Rectangle reaches north and south more than a thousand miles. The broken line, drawn through Manicouagan, divides the larger figure into a smaller Golden Rectangle on top and a bottom square. And in Figure 14, a transatlantic Golden Rectangle (shaded area) and a square join known meteorite locations in Canada, Greenland, and Europe. (Manicouagan-Lappajarvi line served as a guide only, and does not participate in their figures.) Geometries seem to ignore the map's distortion which produces an oversized Greenland.

Figure 15 joins four more local meteorite sites in a search for hidden geometries. They are the Barringer and Odessa craters, first astroblemes to be identified on Earth; plus 1970 Lost City meteorite, first ever recovered from a photographed trajectory; and the 1948 Februarian of Norton, Kansas (it actually landed in Nebraska), then the largest stone ever seen to fall. A Barringer-Lost City baseline opens the hunt. A secret symmetry is revealed by a Golden Rectangle reaching nearly 500 miles across the Southwest. Does this figure explain why the Lost City fireball showed a surprising "lateral acceleration"? Was it being maneuvered to hit a preselected target for delivering the Golden Rectangle on a map?

Can these geometries be dismissed by familiar arguments? "The author is finding what he wants to find." Or, "Given a sufficient number of points on a map, anything is possible." (But if the Golden Rectangles did not exist, they could not be drawn.) Or, "Tiny maps excuse errors. A pencil mark can be five miles broad. Larger charts would instantly disown these alignments." (The only accurate check must come from surveying teams and geodetic satellites.) Or, "Different map projections give varying results." (True. An Azimuthal Equidistant Projection centered on the North Pole delivers a Golden Rectangle larger than the one plotted here!) The final option would have some unknown "law" at work, as suggested by the Field Museum in attempting to account for meteorites' drop-zones.

Until that law comes along, however, we face the inexplicable. For as startling as these rectangles' oceanic dimensions is their amazingly elastic age. They not only engulf hefty portions of Europe, the Atlantic, and North America, but stretch backward through geologic epochs. Massive variations in their dates challenge our theory of continental drift and warp our frame of time.

Another tactic for demolishing an unwelcome notion is to weight it with implications until the idea collapses. Accepting in theory that a design exists within these diagrams, we take the next ludicrous step: these target-sites had to be chosen *before* these meteorites landed on Earth; plan had to precede action.

Two European plot-points, Kaalijarvi and Opava, are a relatively young 5,000 and 20,000 years old respectively, while Rieskessel precedes them by some 15 *million* years. The Canadian Manicouagan is perhaps 200 million years ancient. When that asteroid struck, the dinosaurs had yet to arrive, and Europe, Greenland, and North America lay welded into a single land mass.

Two million centuries ago, according to the theory of continental drift, Europe, Greenland, North America and all other land masses were joined within a super-continent called Pangaea. No individual Atlantic or Pacific existed, only a single encircling sea. For reasons not fully understood, Pangaea began fracturing. From Iceland to Antarctica the fissure widened, opening the Atlantic, separating the Americas from Europe and Africa. Evidence of the broken land weld lies hidden beneath the sea in canyons and ranges snaking down the Atlantic, where submarine volcanoes and seaquakes are spawned even today.

But the Manicouagans and Rieskessels are geologic infants alongside the Hudson Bay Arc. Scientists place its origin in the Pre-Cambrian period when our globe was devoid of life—the land, the sea, the air empty of even microscopic animal and plant forms. The Hudson Bay Arc is between 600 million and a billion years old; the positions of the continents then are imponderable.

How could a system of Golden Rectangles be conceptualized and keyed to future geologic events a half-billion years away? Fitted to the fall of meteorites from space when continents would have inched apart to opening an ocean, and when Stone Age man toted a batch of heavy stones to his Opava campsite?

Or could it be that meteorites are actually *responsible* for the continents' present positions? Two scientists at the 1977 Geological Society of America meeting in Seattle, Washington, suggested that during the past 2.6 billion years, asteroids have periodically blasted loose "great thicknesses of crustal . . . material by cratering." Up through the bottom of these deeply fissured basins burst lava from the Earth's interior in quantities to create mountain ranges. Adult asteroids, acting like a giant punchpress, also may have split the very continents, setting segments adrift upon the molten sea beneath.

But then, why, in all recorded history, has no similar space-bomb landed among us? During this era the human population has been tickled by "babies" like Sikhote-Alin, Siberia; Kirin City, China; Norton, Kansas—not decimated by the likes of a Barringer. Other than nightmare legend, human history offers no account of a direct hit by the space-projectiles that carved out huge craters like Vredefort, Siljan, Lappajarvi, Rieskessel, Sudbury, Manicouagan, Ungava Bay, Gulf of St. Lawrence, Hudson Bay Arc. No holocaust encounter with a mass 10 miles wide and impacting at 35,000 miles per hour, as perhaps excavated Clavius. Not yet.

For Earth's reprieve lies in an incredible streak of good luck. Only by grace is our planet being spared a strike by a Clavius-type asteroid. We came perilously close in 1937, with the near-miss of Hermes.

Chapter Twelve
OUTLAW ASTEROIDS
Planet-X ☆ Space-Bombs ☆ The Heimaey Incident

Our solar system is missing a planet.

According to the chemical testimony of meteorites, according to a mathematical formula known as Bode's Law, and according to the opinion of many scientists worldwide, an Earth-like body, rather than a scattering of little space-rocks, should be orbiting the Sun between Mars and faraway Jupiter.

Theories about a missing planet are nothing new to science: they hark back to the invention of the telescope and the formulation of planetary laws by the noted German astronomer Johannes Kepler (1571–1630). So disruptive was the space-gap to Kepler's mathematics that, with God-like aplomb, "Between Jupiter and Mars I place a planet," he wrote, and continued his historic work on orbits.

Teased for nearly 2 more centuries by this mathematical rip in the fabric of the solar system, astronomers became sorely vexed with the announcement of Bode's Law. This simple but uncanny formula was earlier suggested by J. D. Titius in 1741, and refined by J. E. Bode of the Berlin Observatory. Bode's Law disclosed that the known planets were not haphazardly positioned around the Sun but spaced at rhythmic intervals, and that 3 planets were missing.

Hardly had scientists absorbed this challenging news when another German astronomer, William Herschel, made his momentous 1781 discovery. In all recorded history, Herschel was the first man to sight a new planet in the solar system. He named it Uranus. Positioned on the far side of Saturn, Uranus was spotted almost exactly where Bode's Law had predicted it would be. The search was on for the missing world between Mars and Jupiter.

In the fall of 1800, a society of German astronomers mailed invitations to 24 European astronomers, requesting that each search an assigned segment of the heavens. The mails, stagecoach-slow, were further delayed by war. Before his invitation could reach Italy, a Sicilian monk, Giuseppe Piazzi, saw in his telescope on New Year's Eve a faint, star-like object—one unmarked on his star charts. Piazzi checked its coordinates for several nights and realized that its position was shifting. He assumed this "wanderer" to be a comet, a passing visitor to our solar system. The monk was prevented from pursuing his discovery by illness and cloudy skies.

Calculations by yet another German mathematician, Karl Friedrich Gauss, revealed that Piazzi's "comet" was orbiting the Sun in the great gap between Mars and Jupiter. Despite its midget size, less than 500 miles across, the little body was hailed as the "missing planet." Piazzi named it Ceres after Sicily's patron saint and ancient Rome's harvest goddess. By his discovery the Italian monk had unknowingly opened a new window on our solar system.

On New Year's Eve, exactly one year later, German astronomer Wilhelm Olbers relocated Ceres and, in further searching, found a second "planet" which he named Pallas. Soon to follow were Juno and Vesta. A bevy of sightings during the remainder of the nineteenth century established existence of the asteroid belt.

Then in 1898, startling news about the asteroids rocked astronomy, and implications of that discovery remain—well, Earth-shaking.

During the years just prior to 1898, the finding of yet another little planetoid orbiting the Sun between Mars and Jupiter impressed few scientists. Long past was the heady excitement of 1800–1807, when German and Italian astronomers announced the presence of the first 4 midget planets. In the ensuing decades, amateur sky-watchers had been reporting asteroids in batches. By 1890, star catalogues were clogged by some 300 additions.

With the advent of astronomical cameras in 1890, no longer was there any need to eyeball the object: the technique for identifying asteroids was ingeniously simple. A camera was attached to a telescope geared to track the gradual shift of the stars across a selected portion of the sky. This imperceptibly slow panning action "froze" the stars on the film emulsion into a myriad of neat, white dots. Against this stationary spangle of pinpoints and halos, the reflection of an orbiting asteroid showed as a tiny straight line, its length depending on exposure time, speed, and proximity of the object to Earth. Discoveries poured from the heavens; one observer counted 228.

The technique proved of enormous value to asteroid-hunters, but a nuisance to anyone focusing on more distant reaches of the galaxy. Streaks and scratches were so messing up photographic plates that other astronomers sneered at asteroids as "vermin of the skies" and "rubble of the solar system."

Abruptly, scorn flipped to fervor following the events of August 13, 1898, at Berlin's Urania Observatory. There, a young assistant, Felix Linke, was attending the telescope that routinely photographed the familiar asteroid belt. Exposing the film for two hours, Linke developed it the next morning and noticed a small, unexpected line a fraction of an inch long (.4 mm). Too clean for a flaw in the emulsion, the white snip evidently represented some fast-moving body. A comet, perhaps?

Other sightings allowed mathematicians to calculate an orbit. Then came the surprise of the century.

Unlike the hundreds of its predecessors confined to the asteroid belt 200 million miles and more from Earth, this strange body was traveling cross-country, trespassing onto *our* side of Mars's orbit. Instead of hugging the rail with other asteroids, this steed was leaping the fence, galloping across the centerfield, then jumping the opposite rail to regain the course. The shortcut was carrying the asteroid toward the center of the solar system, and within 14 million miles of Earth.

There was nothing *wrong* with this cosmic steeplechase. The rogue asteroid was trampling no laws of celestial mechanics, nor was it threatening collision with us. Its behavior was just totally unexpected.

In keeping with the venerable and ancient marriage between astronomy and religion, Linke assigned the maverick the name of a male Greek god, Eros, to distinguish it from the more well-behaved "ladies" like Ceres and Vesta. Eros was later established to be a rough slab about 18 miles long, 9 miles wide, and 4 miles thick. Although nowhere near the girth of a midget planet, Eros still looms immense among maverick asteroids. Nevertheless, its presence had gone undetected in 17 photographs taken earlier by other observatories about the globe.

Thirteen years passed before the discovery of another space-object coursing close to Planet Earth. In October 1911, at Vienna's Imperial Observatory (then the world's largest), Johann Palisa observed a faint, fast-moving body stealing inside the orbit of Mars. Distance from Earth: 20 million miles. Palisa bestowed on the tiny rogue the very un-Greek name of Albert, honoring Baron Albert Rothschild, a benefactor to the observatory. Albert winked in and out of our heavens so quickly that no orbit could be traced for predicting its future return to our solar neighborhood. Measuring only about 2 miles wide, little Albert may have come and gone unseen over the years.

Two decades passed before the third male asteroid was spotted on March 13, 1932, by Eugène Delporte of Belgium. He called it Amor. Distance: a respectful 10 million miles from Earth. A mere half-mile wide, Amor taxes even today's telescopes and cameras, remaining the faintest asteroid officially listed.

One third of a century now had elapsed since the initial fervor of Eros: the scene was set for an even more astonishing discovery. Just 6 weeks after Amor, a new breed of asteroid vaulted onto the cosmic stage, narrowing the space-gap between Earth significantly.

The night of April 24, 1932, at a mountaintop observatory known as Konigstuhl ("King's Chair") in the Rhine Valley, Dr. Karl Reinmuth of Heidelberg University was systematically hunting for asteroids. Inspecting his photographic plates, he was startled to find the track of an asteroid that had not only crossed the orbit of Mars, but continued toward the Sun beyond Earth's own path. Distance from us: 6.5 million miles. He named this intruder into our spatial backyard Apollo, after the Greek Sun-god. Established was a new category of asteroids. Today those fast-moving objects approaching the Sun more closely than Earth are called Apollo asteroids. (To avoid confusion with the Apollo Moon Program, however, they are here referred to as outlaw asteroids.)

Astronomers had only just recovered from the close-approach of outlaw Apollo when, in February 1936, Eugène Delporte of Belgium announced Adonis, an asteroid streaking even closer to us at a distance of 1.3 million miles. In Earthly geography, 1.3 million sounds comfortably remote. But for any asteroid on a collision course with our globe, and moving at perhaps double the speed of Pioneer 10, flight-time would be less than 24 hours. Then, on October 28, 1937, an unforgettable asteroid made its debut into scientific journals and into the consciousness of the world.

From his mountaintop observatory, Dr. Karl Reinmuth, Apollo's discoverer, was astonished to see on his film a telltale streak of unusual length—more than an inch long (27 mm). This indicated an object remarkably close to Earth and traveling at surprisingly high speed. But in which direction was the body moving? Toward or away from Earth? Another observation was needed; but clouds over the Rhine Valley blinded his telescope.

Unknown to Reinmuth, the hurtling object had passed from the dark side of our planet to become invisible in the Sun's daylight glare.

Reinmuth alerted other observatories by telegraph, requesting that they inspect any pictures taken during the latter part of October. By good fortune, astronomers in Germany, United States, and South Africa had also photographed that same portion of the sky for other purposes. Together, they retrieved from their files 7 images, permitting a partial orbit to be calculated.

Results were disturbing. On October 31, unknown to anyone on Earth at the time, our planet had experienced its narrowest escape on record from collision with an asteroid. The event made the front page of the world press. *The New York Times* reported, "The two bodies just scraped past each other during the night, in an astronomical sense." The asteroid, which Reinmuth named Hermes, measured some 2 miles wide and weighed upwards of 2 billion tons. It had missed Earth "with a bare 400,000 miles to spare," only double the distance to the Moon. Later calculations opened the gap to 480,000—but still too close for comfort. Hermes' location in the solar system today is a mystery.

Asked about the consequences of a collision between our planet and Hermes, a South African astronomer gave a wry answer. Obviously alluding to Imperial Japanese invasion troops battering China, and to Hitler's military juggernaut ready to smash Europe, the scientist said that the collision "might conceivably have altered the international situation somewhat."

"The visit of Hermes was a near miss for earth inhabitants," observed astro-geologists Robert Jastrow and Malcolm H. Thompson. "If it ever strikes the earth—as it may in the future—the force of its impact will liberate the energy of one billion hydrogen bombs, and may destroy a substantial fraction of the population of the earth."

Hermes marked a new era in astronomy; a new awareness that our planet is not immune to cosmic catastrophe. Since 1937, scientists no longer dismiss the chances of collision between Earth and an outlaw asteroid as infinitesimally small.

Some Pennsylvania residents still recall the near-escape of Pittsburgh on the evening of June 24, 1938. A fireball streaked above the state trailing a cloud of smoke and dust. Moments later, a thunderous explosion rocked the northern section of the city, setting off rumors that a powder magazine had gone off at nearby West Winfield. A mere 10 ounces of particles were collected from the fall area around Chicora, a mountainous and heavily wooded region some 40 miles north of Pittsburgh. Authorities could not tell the size of the original mass. Aircraft collected sample particles from the dust cloud. From its size and density, they estimated it contained over 500 tons of material. A slightly—very slightly—different angle of approach, and this relatively minor asteroid could have obliterated Pittsburgh, killing a half million people.

Even though their numbers are small and their mass meager beside the midget planets of the asteroid belt, the potential blasting power of an outlaw is horrendous. At some unrecorded time in the past, a baby asteroid, weighing less than 300 thousand tons and loafing along at 35,000 M.P.H., smashed into the desert floor of northern Arizona. Result: a crater 3/4 mile wide, 600 feet deep. In far northern Quebec Province, some 4,000 years ago, a heftier infant asteroid of several million tons, blasted a cavity 2½ miles across. If a similar object struck Manhattan Island, reports *Popular Astronomy* (October 1950), "the city would be torn from the Earth and life would be eliminated within a radius of 100 miles by the shock waves; the crust of the Earth would ripple like pond water. This would equal the blast effect of 10 hydrogen bombs, each 1,000 times the power of the wartime A-bomb."

Babies. Imagine something the size of, say, 100 super-tankers; or perhaps a dozen football stadiums the size of the Rose Bowl. Weight: a half-billion tons or better. Velocity at impact: upwards of 80,000 M.P.H.

Consequences of collision with an outlaw asteroid are known, but no accurate predictions in time are possible. It's anybody's guess when the next adult will rendezvous with the third planet from the Sun. Depending on which expert consulted, you have your choice of reassurances—or alarm:

Once in 2 million years, a meteorite large enough to produce a crater strikes the land surface of Earth.
The Earth is struck an average of 3 times every million years by an asteroid measuring a mile in diameter.
An asteroid capable of wiping out a city hits this planet on an average of once every 50,000 years.

An asteroid is expected to strike our planet about every thousand years. Once a century.

An easy, accurate answer is hard to come by, for two reasons. First, the orbits of adult asteroids are subject to change without notice. Second, no one knows how many objects a half-mile or more wide are out there. (Estimates presently range between 100 and 1,000.) And nobody knows how much stuff is flying past unseen. Telescopes and cameras remain inadequate for picking up objects much below a half-mile wide. Reflecting practically no sunlight, they sneak by invisibly, like bats swooping black from the gloom of darkness. When one slinks close enough to kiss our atmosphere, a fireball bursts against our eyes and ears. Witnesses marvel at the meteor's pyrotechnics; scientists frown at its proximity and its explosive power.

Since 1937, the number of discoveries has increased dramatically— perhaps due to heightened interest of the Space Age, perhaps to technological advances in radar, photomultipliers, telescopes, cameras, film emulsions. Undoubtedly, outlaw asteroids that once passed unseen are now being electronically lassooed. Or perhaps the asteroid population is rising?

On April 7, 1959, a fireball blazed near Pribram, Czechoslovakia, about 35 miles south of Prague, laying down a batch of fragments weighing only 20 pounds. By chance 2 sky cameras had caught the meteor. The resulting orbit revealed that the mass had been a true outlaw asteroid.

The Lost City fireball, photographed January 3, 1970, deposited a small, 23-pound stone, but the original mass undoubtedly weighed many tons. Calculations revealed that the parent asteroid had also been an outlaw, orbiting through the asteroid belt and past Mars for perhaps millions of years, before being drawn into a collision course by Earth's gravitational field.

More detailed information is available about the great meteor of August 10, 1972, first sighted near Salt Lake City. At 2:30 P.M., residents and summer vacationers from Utah to Alberta, Canada, were treated to the blazing spectacle of a meteor burning metallic bluish white like a welder's torch, nearly as bright as the Sun. This close-calling asteroid is perhaps the most lavishly recorded in history! Observed by a visiting astro-physicist from the Harvard-Smithsonian Observatory, photographed in color by still and motion picture cameras, tracked in space by an American satellite equipped with a near-infrared radiometer. (Such a heat-sensing device cannot react to the dead cold stone and steel of an asteroid until it warms to the touch of our atmosphere; by then, of course, the intruder is practically upon us.)

Utah-Alberta fireball statistics: altitude, 35 miles. Velocity, 50,000 M.P.H. Size, about 20 feet wide. Blasting potential: according to *Nature* magazine (February 15, 1974), "The initial kinetic energy was . . . approximately the yield of the nuclear weapons which destroyed Hiroshima and Nagasaki." Providentially, this flying bomb did not impact, but faded into invisibility in mid-sky between Calgary and Edmonton. Authorities mention how the fireball "bounced" off our atmosphere, and "skipped" back into space.

Close-calling fireballs, or "Earth-grazers," all as unpredictable as their larger, more distant outlaw cousins, are no strangers to this planet. For the year 1972, the Smithsonian Institution's Center for Short-Lived Phenomena lists 5 in addition to August 10. An average year!

In commenting on the blasting power of an outlaw asteroid discovered in December 1975, the February 7 issue of *Science News* admits, "If it struck the earth, it would make a crater 20 miles across."

Whatever the cause, discoveries have been curiously clustered in time. The year 1976 was a bumper one, heralding 4 new asteroids. *Science News* flagged attention to them with headlines like "An asteroid comes calling—close," and "Still another near-earth asteroid."

The third photographically documented hit by an outlaw occurred February 5, 1977, in Alberta, about 90 miles east of Edmonton, near the community of Innisfree. Two stations of Canada's Meteorite Observation and Recovery Project, similar to the now defunct Prairie Network that captured the Lost City visitor, photographed the fireball's trajectory, revealing the third outlaw orbit. The Innisfree specimen weighed only 4½ pounds.

At this writing (1979) the tally of outlaws discovered since Apollo in 1932 has risen to 28. And this globe is being hit repeatedly by outlaw asteroids: 3 times in the 18-year period from 1959 to 1977. And science has the pictures to prove it. Fortunately for us all these arrivals have been mere space-gnats. Their small size has been our salvation. During the past 2 centuries, scientists have catalogued some 2,000 meteorite specimens. The largest, about 60 tons, still was too small or too slow to explode on contact.

Perhaps we're overdue for a big one to again scar the face of the earth.

With an innocence rare among scientists, some experts are reassured by the oceans that cover nearly three-fourths of our planet's surface. Calling upon the Theory of Probability, they correctly suggest that three out of four asteroids must impact water, naively believing that there lies safety. To the contrary: a resulting tidal wave or *tsunami* several hundred feet to a mile high, would sluice neighboring coastal cities into oblivion.

Indeed, according to *The Secret of Atlantis* by Otto Heinrich Muck (rhymes with book), a "deviant star"—an outlaw asteroid—perhaps 10,000 years ago obliterated that legendary island whose remains form the present Azores. Muck also suggests that a twin projectile from space gouged from the Caribbean sea bottom the 30,000-foot deep Puerto Rico Trench. However, it is unlikely that the fragile ocean floor could withstand a direct hit. Should the shallow basement split under impact of a billion-hydrogen-bomb asteroid, allowing cold ocean to cascade upon the hellish fires seething beneath the Earth's crust, consequences are almost too terrifying to consider.

During the great controversies and discoveries in the early 1800s, two lines of scientific investigation were converging, forming the shaft and the head of an arrow pointing to the "missing planet" between Mars and Jupiter, and to its possible destruction by cosmic cataclysm. Chladni had suggested that extraterrestrial meteorites could be fragments from a shattered planet. Simultaneously, Dr. Olbers, discoverer of the asteroid Pallas, rhetorically inquired of a colleague, "Did Ceres and Pallas always travel in their current orbits in peaceful proximity, or are both part of the debris of a former and larger planet which exploded in a major catastrophe?"

According to a companion theory, remnants of Planet-X spewed across the solar system, to be swept up by its orbiting neighbors as meteorites. Some authorities believe that these cosmic chunks of stone and iron bring with them chemical clues to the structure of Planet-X. It may have been a twin of our globe. Iron meteorites, for example, are seen by many astro-geologists to mirror the molten core of our own world—the stones resembling much of the outer mantle; the rare stony-irons reflecting the region between.

Until recent decades, however, astronomers envisioned no force powerful enough to pulverize a solid body with the magnitude of a

Mercury, Venus, Earth, or Mars. One theory lay in a titanic collision between Planet-X and a similarly-sized visitor to our solar system. But with odds of a trillion to one against a cosmic body meeting another in the lonely vastness of space, scientists rejected the concept.

Then the Atomic Age made its cataclysmic debut in 1945 with the obliteration of Hiroshima and Nagasaki. A plausible source of planet-wrecking power had been found. The October 1950 issue of *Popular Astronomy* magazine informed its readers of the possibility by quoting Dr. Harrison Brown of the Institute for Nuclear Studies at the University of Chicago: ". . . all the meteorites that have ever struck the Earth, and still keep coming (ranging in size from a few milligrams to possibly millions of tons) are the result of a single cosmic catastrophe millions of years ago, in which a planet roughly the size of Mars, which probably occupied an orbit in the gap between Mars and Jupiter, blew up one day, no one knows why. The meteorites, it appears, are the fission fragments of a cosmic atom-bomb, the debris of a planet with a close family relationship to the Earth."

Other authorities agreed. In 1960, Dr. Alan E. Nourse, an expert on space-astronomy, wrote in *Nine Planets*, "We know that an uncontrolled fusion reaction could conceivably split a planet into fragments and consume a good part of the mass of the planet in the process." According to Nourse, the necessary force would require technical assistance from an advanced civilization: "Such a reaction—an enormous hydrogen-bomb explosion—would not be likely to have occurred spontaneously on any planetary body in our solar system at any point in its evolution. If such an explosion did occur, it must have been set off somehow by an intelligence powerful enough to conceive it and build it."

In 1978, two American journalists, Henry Gris and William Dick, recounted their interviews with a trio of Russian scientists who, poetically, refer to the missing planet as Phaeton, in reference to the lad of mythology who impetuously borrowed the Sun-god's chariot for a day. Phaeton was struck down by a Jovian thunderbolt as he recklessly threatened to set Earth afire. According to *The New Soviet Psychic Discoveries*, Dr. Felix Zigel, Professor of Cosmology at the Moscow Aviation Institute, dates the blowup of Phaeton or Planet-X much more recently than Dr. Brown: perhaps only 500,000 years ago.

In reconstructing their Phaeton, Soviets explained to the journalists,

" 'The missing planet . . . had all the external characteristics of our Earth, with oceans, mountains, and an envelope of atmosphere.' " This particular group of Russian scientists suggest that Phaeton, like Planet Earth, was inhabited.

What triggered the planet-wrecking blast? In his discussion of the fate of Planet Phaeton, Dr. Zigel insisted that it "appears probable that it was a thermonuclear explosion, because nothing short of that could have produced so disastrous an effect. The triggering force was man-humanoid. There is no other explanation."

Ah, but there is! If Planet-X was truly Earth-like, with a core of molten iron, sheltered by a thin crust in turn covered by millions of square miles of ocean, then it would be susceptible to a destructive force other than an H-bomb. Let the planet's sea floor be broached, let those global seas pour upon that vast interior furnace, and a doomsday steam explosion could conceivably result.

How might the ocean bottom become unzipped? A mighty space-projectile—an outlaw asteroid—could be the trigger needed. On a smaller, handier scale, a few sticks of dynamite planted at the wrong place at the wrong time on the surface of Phaeton could have achieved a similar result.

After all, our own all-too-fragile globe nearly experienced that ultimate folly in the spring of 1973, when a very small island (Heimaey), owned by a very small nation (Iceland), was almost erased from the map by accident.

Although Heimaey is only a pinpoint on a chart of the North Atlantic Ocean, the world would have noted its unnecessary demise with alarm. Not just because of the cataclysmic power of the "bomb" responsible, but its devastating simplicity.

The explosive force of the "device" ready and waiting was nothing nuclear, yet its blasting potential approached 4 megatons, the equivalent of 4 million tons of TNT. Neither was it the threat of Heimaey's erupting volcano: In its history, all Iceland has withstood hundreds of lava flows without facing the horrifying combination of elements now being innocently prepared by the Coast Guard with an assist from the United States Navy. The explosive components were fire and water: incandescent lavas vomiting from the volcano, and the chilled waters of the Atlantic.

Countdown for Heimaey began in the dark and quiet hours of early morning, shortly after 1:00 A.M., January 23. A workman coming off the night shift at a fish-processing plant on the harbor saw sparks spitting from a field near the edge of town. Beyond lay the dark dome of Mt. Helgafell, "extinct" for some 5,000 years.

Within moments, sparks blossomed into glowing red fountains of lava. Fountains gushed into geysers 400 feet high. The volcano was erupting, not 500 yards from the island's only settlement, Vestmannaey-jar (pronounced Vest-man-ayer). The workman rushed to the fire station to sound the alarm. Orders were given to evacuate the 5,000 inhabitants to Iceland across the narrow strait.

There was no panic. Since the first Viking settlers arrived a thousand years ago, Icelanders have lived with the hot breath of volcanoes breathing down their backs. Soon an armada of fishing boats, coast guard cutters, and pleasure craft was assembled.

Already a downpour of hot ashes was burying the town's deserted streets. Roofs collapsed under the tonnage of cinders. Lava "bombs"—glowing chunks of magma, some the size of basketballs—arched across the sky like incendiary shells, setting homes and businesses afire. Behind the bombs surged the lava: a flaming black cliff splotched with blood-red flame. Steaming, hissing, growling, moving, the advancing wall set houses afire, then bulldozed the embers. Within a week some 100 buildings were engulfed. By the end of May, the toll had reached 360.

Its people safe on the main island, Heimaey's most fearsome threat was no longer loss of human life or destruction of personal and business property. Even the lava-bombed fish-freezing plant on the dock could be replaced. No, Heimaey was staring at economic disaster.

Fishing was the community's sole livelihood, made possible by the small port that provided the island's only berth for boats to bring in their valuable cargoes, one-fifth of Iceland's total catch. A natural break-water of 900-foot cliffs protected the harbor from vicious Atlantic storms. A narrow channel, less than a half-mile wide, offered access to the sea. And now a broad tongue of lava had reached the shore, about to spread across the harbor floor. Should the channel become sealed by the lava flow, Heimaey's only port would become a useless walled-in lake.

The advance of the molten rock was marked on the water's surface by "rooster tail" explosions. Temperature of the usually cold harbor had risen to a tepid 110°F. Already sections of the port were too shallow for the boats. The port must be saved or the fishing community would die financially.

Teams of lava-fighters unmuzzled hoses connected to high-pressure pumps, spraying nearly 30,000 tons of water an hour against the cinder monster. The strategy was not to drown the magma's fires—an impossibility—but only to dampen the lava's leading edge, allowing a crust to form on the broad lumpy face. Cooled, the advancing wall would slow, damming itself, and diverting itself away from the vulnerable harbor. But time had run out for Heimaey. On the harbor bottom, lava had piled nearly 150 feet thick, narrowing the sea channel to a mere 700 feet.

Science had an answer. Geologists would save the port by quenching the fiery advance with the harbor's own waters!

As the molten rock slides into the sea, a crust quickly forms atop the flow, preventing any explosive mixture of heat and cold. The insulating skin shields the main body of lava, while allowing it to creep onward. The geologists would attack the advance beneath the harbor with high explosives, ripping the roof off the lava flow, letting the sea solidify the invader.

The scientists approached the town fathers of Heimaey with the idea of blowing away the insulating skin with dynamite. The fathers approved; the Government of Iceland approved. The Coast Guard, which possessed too small an arsenal for the job, called upon the United States Navy, which flew a team of underwater demolition experts to the aid of embattled Heimaey. The counterattack called for explosives to be planted along the face of the menace one day, and detonated the next. All systems were go!

With detonation only 24 hours away, 2 other scientists made some calculations. One was a professor from the University of Iceland; the other an astronomer, an authority on exploding stars—or supernovae. They sensed a cataclysm in the making.

Even when hot lava and cold ocean blend gradually, there is much boiling of water and hissing of steam. Accelerate the mixture and an explosive phenomenon occurs. The professor and the astronomer reported in *Nature* magazine (August 31, 1973), that "A more catastrophic release of energy is conceivable" should a major lava flow be allowed to mix with ocean water.

They foresaw a chain reaction. The dynamite would rip free the lava's protective roof. The instant cold seawater gushed upon lava hot enough to melt iron, however, an explosion of high-pressure steam of appalling ferocity would result. As great chunks of the insulating skin were ripped loose from the 150-foot thick lava river, the ocean would penetrate to its molten heart. Within minutes—or seconds—the entire lava flow in the harbor would go off like a hydrogen bomb.

How big a blast? The professor and the astronomer were quite specific, estimating the release of energy as between 2 and 4 megatons, the equivalent of 2–4 million tons of TNT—100 times more powerful than the combined yield of the two atomic bombs that pulverized Hiroshima and Nagasaki. Heimaey's 6 square miles of green plains, rolling hills, homes, and fish factories would go sky-high; and with them, undoubtedly, a good chunk of neighboring Iceland. And after the detonation: a tidal wave racing across the North Atlantic, threatening cities from Bergen, Norway, to Manhattan.

No explosives were planted. Heimaey remains intact. The volcano quieted of its own accord, sparing the harbor and the fishing industry. The island community continues to prosper.

The Heimaey incident dramatizes our globe's vulnerability—given the proper trigger—to a planet-wrecking explosion. Although oceans cover nearly three-fourths the world's surface, the term "Water Planet" is a misnomer. As testified by Mt. Helgafell and by scores of other active volcanoes siphoning lavas from that inexhaustible 8,000-mile magma reservoir, our world is essentially a ball of fire, sheathed in a relatively thin crust.

Upon this skin repose the continents, with their human inhabitants plus some 300 million cubic miles of sea water. The crustal separation between water and fire is less than 10 miles—and flawed, at that. Like the seam on a baseball, a fissure twists through the world's underwater basement for 40,000 miles. Lavas have been oozing up through this recently discovered crack for millions of years, giving gradual birth to volcanic islands like the Hawaiian Archipelago in the Pacific; to Tristan da Cunha; the Azores; and Iceland in the Atlantic.

Fortunately, submarine eruptions remain tiny and spotty, mere pinholes in a blast furnace. The enveloping sea easily cools the intruding magmas, permitting them to congeal slowly, gently, as witnessed at Heimaey. Occasionally, though, fire and water mix convulsively, and a Krakatoa occurs.

Krakatoa's "bang heard round the world" resulted when millions of gallons of seawater penetrated the volcano's lava caldron. In a series of detonations the 18-square-mile island in Sunda Strait between Sumatra and Java virtually disappeared, leaving only a broken peak above the water, a 6-mile-wide, 1000-foot-deep basin below. Debris jetted 25 miles into the sky. Atmospheric winds carried the dust around the globe, creating spectacular "Chelsea sunsets" as far away as England. Numerous casualties resulted from the fallout of burning ashes and cinders on neighboring islands, but the true horror arrived with the *tsunami*, or tidal wave. Uplifted to a height of 125 feet in places, the avalanche of water swept away hundreds of villages, drowning an estimated 36,000 people.

Still, the Krakatoa blast must be rated as puny, a local disturbance, nothing global. But should ocean floor and furnace lid ever be broached, our globe could suffer the fate suggested for Planet-X and whose entrails perhaps form the asteroid belt.

Does the necessary force exist for shattering the 40,000-mile seam? Obviously, yes. An asteroid could provide the starter. Two-billion-ton Hermes, which buzzed us in 1937, could have done the job by landing between Labrador and the British Isles, directly above the fracture line known as the Mid-Atlantic Ridge. Here the ocean shallows to 10,000 feet: underwater heights flanking an abyssal gorge, a mile deep and up to thirty wide, or nearly twice the Grand Canyon's breadth. The bottom is a dam between ocean waters and hidden lava—except for localized eruptions, a barrier unbroached these millions of years.

In Greek mythology, Hermes was depicted with wings on his helmet and wings on his heels, for Hermes served as Messenger of the Gods. An extraordinarily fast asteroid, Hermes can be assigned a speed of 75,000 M.P.H.—meaning that following the moment of first sighting by telescope, as in 1937, Earth's population has about 6 hours' reaction time. There is no protection *against* asteroids vectoring for a direct hit. The instant Hermes makes its flaming debut into our atmosphere, only 10 seconds remain.

Beneath the spearing sky-fire, surface waters evaporate, billowing into tempestuous steam clouds. At the instant of impact, the sea shatters like crystal; the shock wave crumpling ships' steel hulls hundreds of miles distant. Unlike mere rock, water is virtually incompressible, offering no "give." Rather than drilling into this barrier, like the Barringer asteroid through Arizona limestone, the projectile vaporizes totally. Temperatures reach millions of degrees. Horrendous pressures shove aside the ocean, opening a bowl twenty miles wide. Tsunami walls of water reach skyward 10,000 feet, then flee toward North America and Europe to sluice into oblivion coastal and lowland cities—time permitting!

Beneath the point-source explosion, a 2-mile column of water vaporizes, baring the sea-floor canyon that is our planet's jugular vein. The blast is a wedge, driving 100 miles down through the Earth's crust and into the seething-hot mantle. As the canyon splits like a ripe melon under an ax, lava vomits into the vacuum above. Towering walls of water collapse back into the empty bowl, red with boiling magma. Krakatoa-like steam explosions feed on each other, widening, deepening, lengthening the lethal wound mile upon infernal mile. Hermes has started an awesome chain reaction, as feared at Heimaey. Earth's thin skin is gashing apart: the 40,000-mile seam becoming unzipped. Unquenchable inner fires flare forth to the attack of 300 million cubic miles of cold ocean.

Prophets of doom have predicted the world will die either by flood or by fire.

They are wrong.

The end comes with a titanic steam bomb.

As reported in *The New Soviet Psychic Discoveries*, the Russians were more accurate in their scenario of Phaeton's demise: " 'The oceans [explode] and with that, the crust of the planet cracks wide open,' Dr. Felix Zigel continued. 'We can assume, therefore, that with its shell destroyed, the planet went on disintegrating until nothing but rubble remained.' "

But Hermes missed us, leaving in its uncharted wake only an uneasy memory among astronomers. Bravado may argue that a miss is as good as a mile, but this ignites the question: why has a doomsday blast never occurred? Because global oceans offer a target 3 times broader than the land masses, theoretically most megameteors would strike an ocean. Yet observe how astroblemes bracket the North Atlantic! How has our Water Planet been spared these millions of years from meeting ocean-busting asteroids like Hermes? Why have our land masses, and never our seas, been victim to enormous astroblemes? Perhaps "somebody up there" is watching over us.

It would seem that we have crossed the line here separating the titillating horrors of science fiction from searing realities. According to several Soviet authorities, however, that line was obliterated on June 30, 1908. Evidence of the gigantic blast above Siberia has been reformulated by Atomic Age and Space Age specialists, and their conclusions are perversely comforting.

CONCLUSION: THE SIBERIAN FIX

Not all fireballs depart so amicably as the Utah-Alberta asteroid, or detonate so harmlessly in the atmosphere as the 1938 Chicora visitor that threatened Pittsburgh. June 1908 witnessed what may have been the most horrendous aerial explosion in the history of the human race. That "asteroid"—if indeed it was a natural member of our solar system—detonated at less than 30,000 feet above northern Siberia. Concussion from the exploding fireball flattened the forest below for 30 miles around, knocked plow horses flat 240 miles away, and delivered thunderclaps 550 miles distant.

Across the breadth of subarctic Siberia there stretches the *taiga*, a landscape feature unlabeled on most topographic maps. Trees by the billions—pine, spruce, cedar, larch, birch—cover for hundreds of miles. The taiga's enormity is legendary. In 1890 the noted Russian author Anton Chekhov observed, "The power and the enchantment of the taiga is not in its gigantic trees or in its silence, but in the fact that only migrating birds perhaps know where it ends . . . You come to the top of a hill, you look ahead, and all you see is another wooded hill, and another, and still another, and so on, without end." Chekhov called the taiga "the green monster," much of which remains unexplored today.

Russian scientists probing its dank secrets are given survival training. Travel through the virtually roadless wilderness is a challenge in winter and nearly impossible in summer. Beneath the taiga's soggy surface, the ground never thaws, and permafrost allows no more leakage than an iron bucket. Swales and vales rimmed by hills permit little drainage. Winter snows dissolve into quagmires of go-nowhere creeks and stagnant swamps that breed "by the cubic mile" clouds of voracious Siberian mosquitoes and stinging, sucking gnats. Russian-born author George St. George in *Siberia* calls them "piranhas of the air," adding, "A man, unless properly protected can actually be eaten alive. . . ."

Today, the taiga near the Stony or Middle Tunguska River where the explosion occurred knows few Russian settlements. In 1908, there were none, other than a few scattered log-cabin trading posts dealing in the "soft gold" of furs from ermine, silver fox, northern otter, and earlier the lustrous sable, brought in by furtive "forest children." Then known as Tungus, today as Evenki, these aborigine people of ancient Mongol-Tartar descent were enduringly resourceful and physically sturdy. The Tungus were nomads, herders of reindeers, who dwelt in portable tents or yurta, covered with hides and not unlike the tepees of North American Indians. Cheerful and peaceable—the Tungus had no knowledge of warfare—they were primitively shy, slipping deeper into their forest fastness at the approach of strangers. Shunning civilization, unable to read or write, speaking their unique tongue, only the superstitious forest children knew the secret abode of the Angry Spirit who arrived among them on that cloudless summer morning.

For nineteen years, war, social upheaval, revolution, and lack of communication prevented the Russian Academy of Science at Petrograd (earlier, St. Petersburg; later, Leningrad) from verifying, let alone locating, the rumored site of a fireball that had decimated a forest in the distant Siberian taiga. In 1921 a preliminary party, headed by Leonid A. Kulik, a pioneer in meteoritics, traveled more than 2,000 miles on the Trans-Siberian Railroad to the village of Kansk, but failed to find the site (they were still nearly 400 miles away). However, Kulik returned with tantalizing eyewitness accounts and old newspaper articles.

In 1927, after incredible hardships (he and his party were forced to subsist for 9 days on plant stems), Kulik arrived at a spot near the Stony or Middle Tunguska River to stand dumbfounded at the spectacle. Three other expeditions followed, seeking the secret of the destroyed forest. After World War II, Soviet attention temporarily was diverted to the more immediate cosmic event of Sikhote-Alin, some 1,700 miles east, where an asteroid had unloaded 30 tons of cosmic iron on February 12, 1947.

After 50 years of investigation, the taiga gave up no evidence of an impacting iron: no explosion craters, no meteorite remnant, no strewn field of particles. The Tunguska fireball left behind no evidence other than burned and toppled trees. Here summarized are observations pertinent to an alien visit, June 30, 1908.

Place: that lightly populated area north and west of Lake Baikal on the taiga's lower fringe. Time, 7:17 A.M. Weather pleasant: Wind calm, sky clear. Peasants and villagers of Kansk reported seeing "a body shining very brightly—too bright for the naked eye—with a bluish-white light." The object was pipe-shaped or cylindrical. One witness likened it to "a chimney lying on its side." Another mentioned observing "a red flying ball with rainbow bands behind and around it. The ball flew on for three to four seconds and disappeared to the northeast." A curious cloud evidently accompanied the object. "The sky was cloudless except that low down on the horizon . . . a small dark cloud was noticed." It discharged a spark or flame. (UFO enthusiasts suggest the cloud may have concealed a mother ship.)

The object's initial altitude as it penetrated our atmosphere is estimated at 80 miles, traveling at an unknown velocity. Generating violent sonic booms, the object descended to a height of some 3 or 4 miles, leveling off on an almost horizontal trajectory. The body approached from the south, but when about 140 miles from the explosion point, while over Kezhma, it abruptly changed course to the east. Two hundred and fifty miles later, while above Preobrazhenka, it reversed its heading toward the west. It exploded above the taiga at 60° 55′ N., 101° 57′ E.

Numerous witnesses reported a blinding sky-flash. According to E. L. Krinov's *Giant Meteorites*, "Many people distinctly saw that when the flying object touched the horizon, a huge flame shot up that cut the sky in two." Others mentioned an explosion "brighter than the sun . . . a huge tongue of flame." A wave of intense heat followed. Scientists estimate the temperature at the explosion's center at 30 million degrees F. Forty miles away, a worker at the Vanavara trading post related, "High above the forest the whole northern part of the sky appeared to be covered with fire. At that moment I felt great heat as if my shirt had caught fire; this heat came from the north side. I wanted to pull off my shirt and throw it away. . . ."

With the heat came hurricane-like winds. "At that moment," the worker continued, "there was a bang in the sky, and a mighty crash was heard. I was thrown onto the ground . . . and for a moment lost consciousness." One hundred miles away, a farmer paused for a rest in his field, and then, as he tells it, "I heard sudden bangs, as if from gunfire. My horse fell onto its knees. From the north side above the forest a flame shot up. Then I saw that the fir forest had been bent over by the wind and I thought of a hurricane. I seized hold of my plow with both hands, so that it would not be carried away. The wind was so strong that it carried off some of the soil from the surface of the ground, and then the hurricane drove a wall of water up the Angara [River]. I saw it all quite clearly, because my land was on a hillside."

According to Soviet scientists, the blast was equivalent to a 30-megaton nuclear bomb: 30 *million* tons of TNT, as opposed to Hiroshima's 20 *thousand*. Following the flash, the heat, and the gale, witnesses reported a towering column of smoke shaped like a spearhead on top. The "fiery pillar" rose to an estimated altitude of over 60,000 feet.

The blast uprooted and knocked down millions of trees, snapping timbers 2 and 3 feet thick like toothpicks. For a distance of 30 miles the mighty taiga lay flattened. Raging forest fires followed. "A very hilly almost mountainous region stretches away . . . towards the northern horizon," wrote expedition leader Kulik. "The distant hills are covered with a white shroud of snow half a meter [twenty inches] thick. From our observation point no sign of a forest can be seen, for everything has been devastated and burned."

Until the close of World War II, scientific consensus held that a major explosion in central Siberia on June 30, 1908, had been a natural phenomenon such as a meteorite or a comet detonating against the Earth. After the war—or more precisely, after August 6, 1945, when an atomic bomb assassinated Hiroshima—opinion began shifting, particularly among a few bold thinkers to suggest that the taiga and Hiroshima suffered a similar fate.

Advocates of the atom bomb theory cite the high-altitude incandescent light (Japanese called it *pikadon*, "flash-boom"), the lethal heat, the concussion, the winds. Such evidence seems to point to a nuclear device, the only known force capable of such widespread destruction, while pointing *away* from any natural phenomenon like a disintegrating meteorite.

In mid-October 1976, the official Soviet news agency, Tass, released a story to the wire services of the world quoting a prominent scientist, Alexei Zolotov. Following a 17-year investigation into the blast, Zolotov, an authority on geology, physics, and mathematics, proposed that a spaceship, controlled by "beings from other worlds," may have been responsible for the 1908 explosion above the Siberian wilderness. According to Zolotov, "what took place was a nuclear explosion." He estimated its size as between 20 and 40 megatons, or up to 2,000 times more powerful than the Hiroshima A-bomb. He and his colleagues suggested that the spacecraft itself was nuclear-propelled: suffering a malfunction, it had blown up accidentally.

The hypothesis carries little credence, as Zolotov admits: "I cannot see that a civilization capable of constructing such a ship, piloting it through space to Earth, would not have sufficient safety factors built in to prevent this happening." Indeed, also arguing against a nuclear flareup is the bomb's benign nature. Where Hiroshima mourned nearly 100,000 dead, Siberian peasants and nomads escaped serious injury. If that "thing" had detonated anywhere near a populated area—disaster! "You could hardly have chosen a better place—so far away from our cities and so empty of life," observes Zolotov, "to explode a nuclear device . . . The Tungusky explosion was an amazing demonstration of pinpoint accuracy and humanitarianism."

The Tunguska blast's timing seems too fortuitous for an accident. Less than a 5-hour delay and our rotating planet would have presented the then-St. Petersburg as a target to the space-object. Three hours earlier, and the powerful projectile could have impacted the Pacific Ocean where abyssal trenches border Siberia, Japan, and the East Indies. Here plunges the Mariannas Deep, part of the fragile seam lacing our planet's crust. The tiniest course change in space, and the 1908 bomb might have exploded among China or India's teeming millions. Can we assume that the "pilot" chose a cloudless day with excellent visibility from aloft to assure a safe drop? American military strategy called for identical weather conditions; for a perfect strike on Hiroshima's industrial heart, the *Enola Gay*'s bombardier was forbidden to release through a cloud cover: he had to see the target below. To maximize blast destruction, minimize radiation perils: the bomb was set to explode at a high altitude, rather than against the ground. Similarly, the Siberian missile detonated high in the air, reducing or even eliminating fallout hazard.

Further argument against any accident is the vehicle's flight path. After plotting the 1908 explosion site at 60° 55' N., 101° 57' E. with key sites on a map of Asia, I submit that the accident-explanation is untenable: the sighted fireball was on no haven-seeking emergency course. Instead, the flaming object was being expertly navigated, its flight path oriented to prominent landmarks.

A map of northeast Asia is skeined with look-alike rivers and ribbed with no-name hills and mountains. Yet upon this vast expanse that is

Siberia, one feature can be distinguished immediately—the slim crescent of Lake Baikal. "A gigantic gash cuts a remote tableland of Central Asia as with a titan's blade," writes Ferdinand C. Lane in *The World's Great Lakes*. "Filled with water, cold and beautifully clear, it reflects the tangled mountain peaks rising to a height of 5,000 feet that surround it on every hand."

Baikal's no lengthier than Lake Michigan (about 400 miles), and averages only half as wide (some 35 miles), but author Lane rightly extolls Baikal as the "Colossus of Fresh Waters" because of its unchallenged depth. Its surface lies 1,500 feet *above* sea level, while its bottom rests four-fifths of a mile *below* sea level. As the world's deepest lake, the Baikal gorge holds some 15,500 *cubic miles* of fresh water, 18 percent of the globe's surface total, or nearly the combined volume of all five Great Lakes. To the Russians, Baikal is the Sacred Sea; to roving Mongols it is *Dalai Nor*, the "Holy Lake."

Geologically, Baikal is a water-filled rift in the Earth's crust, opened by global forces that cause continents and ocean basins to slowly slip about during millions of years. In midsummer, when its icy visage melts at last, Baikal shines an exquisitely deep blue color, an inner radiance issuing from incomparably limpid depths that filter out less brilliant hues. On a rare cloudy day, Baikal may show a grayish cast to shore observers, but aquamarine and even turquoise from any elevation. With its craggy crescent shape, it would have provided an excellent visual aid to an alien pilot. The craft apparently approached its target from the south, coming in over the Himalayas, highest mountains on Earth; above the Gobi Desert; and into Siberia. Lake Baikal would have served as an excellent navigational aid, no less than the Great Lakes for the Canadian Fireball Procession.

Puzzling but pertinent are two abrupt course changes by the vehicle when northwest of Lake Baikal, and above the villages of Kezhma and Preobrazhenka. These two sharp course changes have been interpreted by some authorities as a last-ditch emergency maneuver by the nuclear-propelled vehicle about to explode. The map-diagram (Fig. 16) suggests otherwise.

To minimize chance and coincidence, a "measuring stick" is needed. Instantly available is the Golden Rectangle from Henbury. This eye-pleasing mathematical figure can be applied in testing the spaceship explanation. Five points now are available for map-plotting in search of

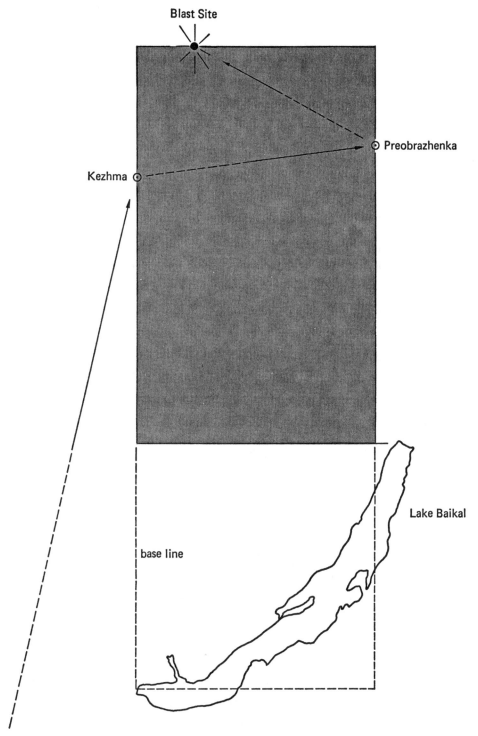

Figure 16. Schematic of the 1908 Blast Site   The cities over which the object changed course, and the site of the final explosion, as plotted in relation to Lake Baikal. (Flight path adapted from *The Fire Came By*)

the Golden Rectangle: they are the two villages, the tips of Lake Baikal, and the blast coordinates of 60° 55′ N., 101° 57′ E.

A Golden Rectangle stands atop a square, formed by lines involving 4 sites: Lake Baikal tips, two villages, and blast point. These geometries suggest that the 1908 fireball was following a prescribed flight plan, and engaged in no emergency maneuvers. Map-diagrams also suggest that these subarctic coordinates later served for positioning the fall of two giant meteorites on Asian soil—Sikhote-Alin in 1947 and Kirin City in 1976.

Now that still other Golden Rectangles emerge from the map, what then? Nothing has been *proved*. Evidence will suggest, however, that a marriage has taken place among extraordinarily diverse sites, implying they are not isolated by time and by geography but rather, enjoy a cosmic connection.

Ever since study began in 1927, most scientists have regarded the Siberian explosion as an isolated event, divorced in time and geography from other terrestrial occurrences. In fact, the entire field of meteorites is pervaded by a similar attitude, unhappily reinforced by this very book. Because the subject of fireballs, meteorites, and star-wounds is so broad, it must be partitioned into chapters. This convenient yet misleading division blocks the mind from seeking associations. Unconsciously, for example, we blank out any connection between the fall of a stone at Lost City, Oklahoma, and the discovery of an astrobleme in Sweden; between a family of giant irons in Greenland and a cluster of stones in Kansas; between a square crater in northern Arizona and a Golden Rectangle near Henbury, Australia. After all, what reason or purpose could conceivably link them together?

Suggestions that Earth has been visited by "ancient astronauts" or is presently under surveillance by UFOs has either been scorned as cheap sensationalism by the scientific community or met with good-natured laughter. And without documentation, scientists have had little alternative.

But have today's experts inherited the Age of Reason's cold conceit that scorned as fable reports of stones dropping from the skies? Absurd at first thought, the notion of cosmic control over incoming fireballs and meteorites does not sound so laughable now that the evidence is in. Once we admit that this mathematical finesse is neither natural nor coincidental, aren't we facing the unthinkable explanation? Unseen and unsuspected by Earthlings, an intellectual force is busily at work vectoring asteroids to preselected target sites and sculpting the third planet from the Sun to its liking.

But is the idea of cosmic inhabitants so idiotic to "primitive" Bedouins conspiring to withhold information about their "stone that fell from heaven"? Or to Greenland Eskimos who undoubtedly know the whereabouts of the Man, expelled from heaven by the Evil Spirit? Or to Indians of our Northwest, who venerated their Moon Stone? Or to the hosts of untutored people who, since remote antiquity, have recognized meteorites as "Messengers of the Gods"? Finally, is the idea so inane to the Evenki of Siberia who in 1908 witnessed the sky-searing aerial maneuvers of something "unthinkable" from outer space?

The existence of extraterrestrial beings, operating somewhere in the vicinity of our Sun system, would resolve many of the puzzlements met in *Stones From the Stars*. We need only to ascribe to them an advanced technology that permits them to manipulate the flight paths of asteroids, fireballs, and meteorites. The 1970 Lost City, Oklahoma, fireball, which opens this book, and the 1908 Siberian space-visitor, which closes it, are as good illustrations as any.

Scheduled Earth-shots by alien intellects would account for timetables like the 3:00 P.M. Peak and the host of Februarians like Sikhote-Alin, Siberia; Kirin City, China; and Allende, Mexico. Their preselected drop-zones would solve the mystery of the "geography of meteorites" as witnessed by Iron Alley, the Hermit Kings, and the Kansas collection. A penchant for things geometric would explain Asia's Golden Rectangles, Quebec's great circle, and northern Arizona's Barringer square. Why, then, would anyone want to drop chunks of stone and iron upon our globe? Once a motive is isolated, clues can be assembled to prop up the heady possibility that space-rocks have a mind of their own.

Find a rationale for these affairs, and we edge closer to conviction that they are not born of Mother Nature, and are not the bastard offspring of Chance and Coincidence, but are family members within some kind of cosmic plan.

A further suggestion may lie in Iron Alley. According to Indian legend, a god transported 11-ton El Morito to central Mexico as a boundary marker between newly arrived tribes. Tradition also claimed that the giant iron meteorite could speak to the Indians—though both the tongue used and the message delivered are unknown. Obviously, here we deal in metaphor: a boulder set with a bronze benchmark on a county line can also be said to "talk" to surveyors by telling its latitude and longitude.

For anyone exploring new terrain, maps are of high priority. For example, when our astronauts took "One giant step for mankind" upon the lunar surface, they were guided by highly reliable charts, compiled from photographs taken earlier by terrestrial telescopes and from close-up television images transmitted by unmanned craft before impact. No human footprints have yet tracked the Martian landscape, yet maps detailing its peaks and plains are ready for future treks. Decades more may pass before some astronaut or cosmonaut steps ashore on Sun-broiled Mercury, yet radar and television have already tentatively charted its cratered face.

The assumption seems safe that exploring parties from other civilizations, venturing into our solar system, also would require maps of the planets—including the pretty blue one so covered with water. Ideally, a region is first surveyed on the ground, and a system of markers are positioned precisely in terms of latitude and longitude. These bench-marks serve as a framework for later mapping operations.

Are meteorites and craters performing a similar function for alien surveyors?

Further, the humanitarianism of the 1908 bomb, the benign be-havior of incoming asteroids (Hermes missed us), and hazard-free meteorite falls also imply a protective policy toward mankind. Who knows how a trusting, healthy human race, forever eager to serve its unseen gods, fits into alien plans? For indeed, meteorite identification and meteorite worship are not confined to "rude" cultures like Eskimos and Indians. Any probings into the history of today's great religions—Christianity, Islam, Buddhism, Judaism—will find enshrined a meteorite. Rome, Jerusalem, Thebes, Delphi, Ephesus, Mecca: all in their day venerated a "sacred stone."

The hidden significance of cartography in the affairs of man (and his gods) is too vast a subject for further investigation here. Another book is needed to plumb the secret union between religion, maps, and meteorites.

BIBLIOGRAPHY

Alderman, Richard Arthur "Meteorite Craters at Henbury, Central Australia." *Mineralogical Magazine*, XXIII, No. 136 (1932).
Asimov, Isaac *The Solar System and Back*. New York: Doubleday, 1970.

Baldwin, Ralph B. *The Measure of the Moon*. University of Chicago Press, 1963.
Barringer, Brandon "Historical Notes on the Odessa Meteorite Crater." *Meteoritics*, Vol. 3, No. 4. December 1963.
———. "Daniel Moreau Barringer and His Crater." *Meteoritics*; Vol. 2, 1964.
Barringer, Daniel Moreau "The Meteorite Search." *Natural History*, May 1964.
Baxter, John and Thomas Atkins *The Fire Came By*. New York: Doubleday, 1976.
Becker, John V. "Re-entry from Space." *Scientific American*, January 1961.
Boggild, O. B. "The Meteoric Iron from Savik Near Cape York, North-Greenland." *Contributions to Mineralogy, No. 22*: University of Copenhagen, 1927.
Branley, Franklyn M. *Comets, Meteoroids, and Asteroids*. New York: Thomas Y. Crowell, 1974.
Brown, Peter Lancaster *Comets, Meteorites & Men*. New York: Taplinger, 1974.
Burkard, Capt. R. K. *Geodesy for the Layman*. St. Louis: Aeronautical Chart and Information Center, February 1968. (Pamphlet)

Caidin, Martin "Mountains from Space." *Argosy*, February 1974.
Cassidy, William A. "Meteorite Field Studies at Campo del Cielo." *Sky and Telescope*, July 1967.
Cassidy, William A., Luisa M. Villar, Theodore E. Bunch, Truman P. Kohman, Daniel J. Milton, "Meteorites and Craters of Campo del Cielo, Argentina." *Science*, September 1965.
Chapman, Clark R. "The Nature of Asteroids." *Scientific American*, January 1975.
———. *The Inner Planets*. New York: Scribners, 1977.
Charters, A. C. "High Speed Impact." *Scientific American*, November 1970.
Clarke, Roy S. Jr., Eugene Jarosewich, Brian Mason, Joseph Nelson, Manuel Gomez, Jack R. Hyde, "The Allende, Mexico, Meteorite Shower." *Contributions to the Earth Sciences*; No. 5, Smithsonian Institution Press, 1970.
Corliss, William R. *Handbook of Unusual Natural Phenomena*. Glen Arm, Maryland: The Sourcebook Project, 1977.

Cornell, James *It Happened Last Year! Earth Events–1973*. New York: Macmillan, 1974.

Dietz, Robert S. "Astroblemes." *Scientific American*, August 1961.

Farrington, Oliver C. *Meteorites*. Privately printed by author, 1915.
———. *A Century of the Study of Meteorites*. Smithsonian Annual Report, 1901.
———. *Catalogue of the Meteorites of North America*. National Academy of Science, 1909.
Firsoff, V. A. *Exploring the Planets*. New York: Barnes, 1964.
Florensky, Kirill P. "Did a Comet Collide with the Earth in 1908?" *Sky and Telescope*, November 1963.
Foster, George *The Meteor Crater Story*. Winslow, Arizona: Meteor Crater Enterprises, 1964.

Green, Fitzhugh *Peary, the Man Who Refused to Fail*. New York: Putnam, 1926.
Greenwood, David *Mapping*. University of Chicago Press, 1964.

Henderson, E. P. and Hollis M. Dole "The Port Orford Meteorite." Oregon Historical Society, *The Ore Bin*, July 1964.
Hobbs, William H. *Peary*. New York: Macmillan, 1937.
Heide, Fritz *Meteorites*. Translated by Edward Anders. University of Chicago Press, 1964.

Jacchia, Luigi G. "A Meteorite that Missed the Earth." *Sky and Telescope*, July 1974.

Kaiser, T. R. *Meteors*. New York: Pergamon Press, 1955.
King, Elbert A. *Space Geology*. New York: John Wiley, 1976.
Krinov, E. L. "Fragmentation of the Sikhote-Alin Meteoritic Body." *Meteoritics*, September 30, 1974.
———. *Giant Meteorites*. Translated by J. S. Romankiewicz. New York: Pergamon, 1966.
———. "New Studies of the Sikhote-Alin Meteorite Shower." *Sky and Telescope*, February 1969.
———. *Principles of Meteoritics*. Translated by Irene Vidziunas. New York: Pergamon, 1960.
———. "The Siberian Meteorite Fall of February, 1947." *Sky and Telescope*, May 1956.

Lamar, Donald L. and Mary Fromic "Electromagnetic Effects Associated with the San Francisco Fireball of November 7, 1963. *Meteoritics*, February 1964.
Lange, Erwin F. "Dr. John Evans, U.S. Geologist to the Oregon and Washington Territories." *Proceedings of the American Philosophical Society*, June 15, 1959.
LaPaz, Lincoln "Hunting Meteorites: Their Recovery, Use, and Abuse from

Paleolithic to Present." *Topics in Meteoritics, No. 6*, University of New Mexico Press, 1969.

Lawrence, William L. *Dawn Over Zero*. New York: Knopf, 1946.

Lehner, Ernst and Johanna *Lore and Lure of Outer Space*. New York: Tudor, 1964.

Leonard, Frederick C. "A Fifth Member of the Cape York Siderite Fall." *Meteoritics*, Vol. 1, No. 3, 1955.

Leonard, George H. *Somebody Else Is On the Moon*. New York: David McKay, 1976.

Ley, Willy *Watchers of the Sky*. New York: Viking, 1969.

Mason, Brian *Meteorites*. New York: John Wiley, 1962.

McCall, G. J. H. *Meteorites and their Origins*. New York: John Wiley, 1973.

McCrosky, Richard E. "The Lost City Meteorite Fall." *Sky and Telescope*, March 1970.

Mebane, Alexander D. "Observations of the Great Fireball Procession of 1913 February 9, Made in the United States." *Meteoritics*, Vol. 1, No. 4, 1956.

Milton, Daniel J. "Structural Geology of the Henbury Meteorite Craters, Northern Territory, Australia." Geological Survey Professional Paper 599-c. U.S. Government Printing Office, 1968.

Monod, Theodore "The Problem of the Chinguetti French West Africa Meteorite," *Meteoritics*, Vol. 1, No. 3, 1955.

Muck, Otto *The Secret of Atlantis*. New York: Quadrangle, 1978.

Nininger, Harvey H. *Our Stone-Pelted Planet*. Boston: Houghton Mifflin, 1933.

———. *Out of the Sky*. University of Denver Press, 1952.

———. *A Comet Strikes the Earth*. Denver: American Meteorite Laboratory, 1969. (Pamphlet)

———. *Find a Falling Star*. New York: Paul S. Eriksson, 1972.

Nourse, Alan E. *Nine Planets*. New York: Harper, 1960.

O'Keefe, John A. "Tectites and the Cyrillid Shower,"*Sky and Telescope*, January 1961.

Peary, Robert E. *Northward Over the "Great Ice."* New York: Frederick A. Stokes, 1898.

Philby, H. St. John. "Rub' Al Khālī: An Account of Exploration in the Great South Desert of Arabia." *Royal Geographical Journal*, January 1933.

Rinehart, John S. "Some Observations on High Speed Impact." *Popular Astronomy*, November 1950.

Robinson, Arthur H. and Randall D. Sale *Elements of Cartography*. New York: John Wiley, 1969.

Sagan, Carl *The Cosmic Connection*. New York: Anchor Press, Doubleday, 1973.

Shepard, Jean and Daniel *Earth Watch*. New York: Doubleday, 1973.

Strahler, Arthur N. *Physical Geography*. New York: John Wiley, 1960.

Watson, Fletcher G. *Between the Planets*. Cambridge: Harvard University Press, 1956.
Weems, John Edward *Peary, the Explorer and the Man*. Boston: Houghton Mifflin, 1967.

INDEX

In the 18th century, French scientists denied that stones could possibly fall from the sky. Yet modern science still tries to ignore meteorites' baffling behavior. They defy the laws of physics by slowing down, speeding up, even making 180° course changes. And instead of landing randomly about the globe, they favor certain drop-zones. Appalachia, Kansas, and the narrow "Iron Alley" of America's West Coast have received repeated bombardments over the centuries, and other regions none at all!

For nearly five years, T.R. LeMaire re-examined astronomical journals, searched explorers' accounts, and correlated the scientific literature on meteors to reveal a plethora of startling anomalies. Why do "Sociable Stone" meteorites usually fall near human habitation, while "Shy Irons" descend in remote desert regions? (The only meteorite known to strike a human being was a Sociable Stone; what may be the largest iron on earth—the size of a 12-story apartment house—is also the shyest of all, having been lost in the Sahara since its discoverer first glimpsed it 60 years ago.) When modern experts find it hard to spot authentic meteorites, how did the world's "primitive" cultures so easily identify "stones from heaven" that fell centuries before?

Still more puzzling are the precise geometries that meteorite sites reveal. Impact craters here and on the moon